The Long-Term Adequacy of World Timber Supply

The Long-Term Adequacy of World Timber Supply

Roger A. Sedjo
Kenneth S. Lyon

Resources for the Future
Washington, DC

Printed in the United States of America

Published by Resources for the Future
1616 P Street, N.W.; Washington, D.C. 20036

Books from Resources for the Future are distributed worldwide by
The Johns Hopkins University Press.

Library of Congress Cataloging-in-Publication Data
Sedjo, Roger A.
The long-term adequacy of world timber supply / Roger A. Sedjo and Kenneth S. Lyon.
p. cm.
Bibliography: p.
Includes index.
ISBN 0-915707-46-2 (alk. paper)
1. Lumber trade. 2. Forest reserves. I. Lyon, Kenneth S.
II. Title.
HD9750.5.S36 1990 89-8519
333.75′ 11—dc20 CIP

This book is a product of RFF's Energy and Natural Resources Division, Douglas R. Bohi, director. The copy editor was Martha S. Cooley; the project editor, Dorothy Sawicki. The book was designed by Joan Engelhardt. The index was prepared by Margaret A. Lynch.

∞ The paper in this book meets the guidelines for permanence and durability of the Committee on Production Guidelines for Book Longevity of the Council on Library Resources.

Printed on recycled paper.

RESOURCES FOR THE FUTURE (RFF) is an independent nonprofit organization that advances research and public education in the development, conservation, and use of natural resources and in the quality of the environment. Established in 1952 with the cooperation of the Ford Foundation, it is supported by an endowment and by grants from foundations, government agencies, and corporations. Grants are accepted on the condition that RFF is solely responsible for the conduct of its research and the dissemination of its work to the public. The organization does not perform proprietary research.

RFF research is primarily social scientific, especially economic. It is concerned with the relationship of people to the natural environmental resources of land, water, and air; with the products and services derived from these basic resources; and with the effects of production and consumption on environmental quality and on human health and well-being. Grouped into four units—the Energy and Natural Resources Division, the Quality of the Environment Division, the National Center for Food and Agricultural Policy, and the Center for Risk Management—staff members pursue a wide variety of interests, including forest economics, natural gas policy, multiple use of public lands, mineral economics, air and water pollution, energy and national security, hazardous wastes, the economics of outer space, climate resources, and quantitative risk assessment. Resident staff members conduct most of the organization's work; a few others carry out research elsewhere under grants from RFF.

Resources for the Future takes responsibility for the selection of subjects for study and for the appointment of fellows, as well as for their freedom of inquiry. The views of RFF staff members and the interpretation and conclusions of RFF publications should not be attributed to Resources for the Future, its directors, or its officers. As an organization, RFF does not take positions on laws, policies, or events, nor does it lobby.

Contents

Appendixes

List of Figures and Tables

Figures

Tables

Foreword

The Long-Term Adequacy of World Timber Supply is the culmination of an ambitious effort to assess the long-term availability of the world's timber. This book continues a tradition of research at Resources for the Future (RFF) on the scarcity and future availability of natural resources that dates back to its publication in the early 1960s of *Trends in Natural Resource Commodities* by Potter and Christy; *Scarcity and Growth: The Economics of Natural Resource Availability* by Barnett and Morse; and *Resources in America's Future: Patterns of Requirements and Availabilities, 1960–2000* by Landsberg, Fischman, and Fisher.

Roger A. Sedjo and Kenneth S. Lyon bring a new tool—control theory—to the study of the adequacy of global timber supply. Their choice of methodology is particularly well suited to modeling the long-term transition from a timber supply derived largely from old growth stocks to one that becomes increasingly dependent on plantation forestry. In analyzing the transition, this volume provides both theoretical and empirical justification for challenging the conventional wisdom that real timber prices will rise for the indefinite future.

The task of modeling global timber supply is an enormous one, necessitating some simplifying assumptions to fit the available data and resources. Questions can and undoubtedly will be raised about some of the assumptions and data, and caution is certainly advised in the use of specific projections from the model. Nevertheless, the authors succeeded in providing a reasonable and defensible characterization of the world timber economy. Moreover, the authors' analysis of alternative scenarios offers useful insights into the effects on prices and regional production levels of employing different assumptions regarding the structure of exchange rates, the establishment of new forest plantations, harvest levels in the Soviet Union, and taxation of the timber industry in the United States.

While this study provides many interesting insights, its principal contribution may well lie in its influence on future research. In developing the first economic model of world timber supply, Sedjo and Lyon provide a framework for analyzing the implications of managerial and technological breakthroughs that might increase future timber growth; policy shifts that might alter planting or harvesting rates in principal producing regions; climatic and environmental changes that might alter timber growth in specific regions; or any other adjustments likely to have important effects for the global timber market. In addition, as better data on timber inventories, growth, and production costs are developed and as larger computers and research budgets can be brought to bear on the task, further disaggregation and refinement of this model will be possible.

International forestry issues have been a major area of research at Resources for the Future since Roger Sedjo joined the staff in 1977 as the first director of the Forest Economics and Policy Program. Analysis of the multiple use of public forestlands by Michael D. Bowes and John V. Krutilla has been another important area of inquiry within RFF's forestry program; *Multiple-Use Management: The Economics of Public Forestlands* (1989) draws together much of their work on this subject.

The Forest Service and the Weyerhaeuser Company Foundation have provided major funding for the Forest Economics and Policy Program since its inception.

January 1990

KENNETH D. FREDERICK
Senior Fellow
Energy and Natural
Resources Division
Resources for the Future

Preface and Acknowledgments

This study represents a continuation and logical extension of earlier work undertaken by the Forest Economics and Policy Program of Resources for the Future (RFF). That work dealt with timber supply issues and particularly with the question of the current and likely future role of the United States and North America as producers of industrial wood in the global context. The initial phase of the work involved an investigation of postwar global industrial wood production and trading patterns; it culminated in RFF's publication of *Postwar Trends in U.S. Forest Products Trade* by Sedjo and Radcliffe in 1980 and *Issues in U.S. International Forest Products Trade* by Sedjo in 1981. The second phase of that early work investigated the comparative potential of industrial forest plantations in twelve regions of the globe, including the Pacific Northwest and the U.S. South, and was completed in 1983 with the publication of the Sedjo study on *The Comparative Economics of Plantation Forestry: A Global Assessment.*

The present study, which examines the adequacy of long-term timber supply, is essentially an extension and a broadening of these earlier works. They examined the existing production and trading patterns of industrial wood and the economic viability of investments in industrial forest plantations in numerous regions around the globe on the assumption that these producers would be small and thus unable to influence price in a world with a variety of wood sources. This study examines the potential economic wood supply for the next fifty years from all of the major sources. As is taken into account in this new study, the production of the newly emerging forest plantations is just part of the world's wood basket, which includes old growth forests, secondary growth forests, managed and unmanaged stands, and native and exotic species. In addition, the approach taken here recognizes that new lands may come into forest production while other forestlands may drop from the commercial forest base as prices and costs change. There is also recognition of shifting comparative advantage as the economics of timber growing gradually replaces the economics of old growth harvesting. Further, the approach that we take recognizes that trees are renewable often even without investments in regeneration (typically the forest will return if alternative land uses do not impede natural regeneration, as has been the experience in the Lake States, New England, and much of the South), but that with appropriate investments, the process of timber growing can be dramatically shortened and regeneration can be controlled for quality and desired species. Finally, this study acknowledges the important role that technological change has played in the past and is likely to play in the future as a mechanism for offsetting resources that are in short supply while increasing the use of readily available resources.

The economic availability of a resource, however, does not depend solely on supply. Hence this study investigates factors that drive demand, particularly over time. Of course, it is the interaction of supply and demand that ultimately determines price and harvest levels and the general perception of "timber adequacy."

Numerous individuals and organizations assisted in the completion of the timber supply project and in the preparation of this manuscript. At the project's inception Ross Whaley, then of the U.S. Forest Service and currently president of the College of Environmental Science and Forestry at the State University of New York, provided both initial encouragement and a challenge with the observation that although the control theory approach appeared to have great promise for forestry analysis, it had yet to demonstrate that promise. Early support for the project was provided by the several journals that saw fit to publish a number of our early papers in which the timber supply model was developed and preliminary empirical applications were undertaken.

During the early stages of this project we were privileged to spend two months at the International Institute for Applied Systems Analysis (IIASA) in Laxenburg, Austria, while the theoretical model was being refined and modified to allow its empirical application. The IIASA working environment and support, together with the intellectual stimulation offered by our IIASA colleagues Lars Lonnstedt, Risto Seppala, and others, were invaluable in that early phase of development. Our subsequent association with the IIASA Forest Sector Project provided additional opportunities for a continuing dialogue, and we wish especially to thank Markku Kallio and Dennis P. Dykstra.

Throughout the life of the RFF timber supply project, financial support was provided by the Weyerhaeuser Company Foundation. In addition, intellectual and moral support was provided by John McMahon, David Mumper, R. N. Pierson, and Dale Kalbfleisch of Weyerhaeuser Company. The company's technical staff provided careful readings, comments, and suggestions on numerous portions of the work, from the early conceptual and technical papers to the draft manuscript of this book. Especially helpful were R. N. Pierson, Dale Kalbfleisch, and, in the early stages, William Lange, now with the Forest Products Laboratory of the Forest Service. We must also thank Mary Hall, Susan Raynolds, and Kenneth Miller of the Foundation for their support of our efforts.

A number of other firms in the forest products industry provided both financial and moral support. Especially notable for their unwavering support have been Donald Taylor of Champion International, and William Ticknor and Herbert Winer of the Mead Corporation.

During much of the early development of this project, direct financial support was provided by the U.S. Forest Service. R. Max Petersen, then chief of the Forest Service, and Robert Buckman, then deputy chief for research, were especially helpful. Financial support was also provided by the Utah State University Economics Department and the Agricultural Experiment Station (Project No. UAES-061).

A number of readers reviewed earlier versions of this manuscript. Detailed reviews of the entire manuscript were provided by Clark S. Binkley, Marion Clawson, Edward Dickerhoof, B. Delworth Gardner, Lloyd Irland, Robert Stone, and Herbert Winer. Reviews that were focused more directly upon specific sections included those of Neil Brett-Davies, I. J. Bourke, David Darr, Dennis Dykstra, Hans M. Gregersen, G. Robinson Gregory, Dale Kalbfleisch, Perry Hagenstein, Richard Haynes, Jan Laarman, John McMahon, T. J. Peck, R. N. Pierson, C. F. L. Prins, Kenneth Skog, W. R. J. Sutton, Phillip Wardle, and Herbert Winer. In addition, a number of anonymous referees provided helpful comments and suggestions.

The data demands of this project were substantial. References and data were provided by Roy Beltz and Robert Curtis of the Forest Service, and by Bambang Adiwyioto, who provided data and model adaptation in his dissertation, from which much of the material on the Asia-Pacific region was drawn. Research assistants Katherine Tunis and, later, Ellen Rom assisted in collecting the considerable amounts of data required for the study. In addition, Vivian Papsdorf, staff assistant for the Forest Economics and Policy Program, deserves considerable thanks for her efforts on various draft manuscripts.

We wish to thank Emery N. Castle, former president of RFF, who supported the project from its inception to his retirement, and Kenneth D. Frederick, then director of RFF's Renewable Resources Division, who provided constructive challenges to the project and to the manuscript that ultimately improved the final version. Finally, we wish to thank Robert W. Fri, RFF's president, and John F. Ahearne, RFF's former vice-president, for their support of the project to its completion.

ROGER A. SEDJO
Senior Fellow
Energy and Natural Resources Division
Resources for the Future

KENNETH S. LYON
Professor of Economics
Utah State University

1

Introduction

Is the world running out of usable timber? Or is such a concern alarmist and without substantive merit? The adequacy of timber supply has been a societal concern throughout much of history. As early as the fourth century B.C., the Guanzi, a Chinese legal document, provided government officials with guidance for ensuring a long-term, continuous supply of timber. By the Middle Ages the discipline of forestry had been developed in Europe, largely as the consequence of inadequacies in supply that resulted from pressures on the natural forest. Thus both the Chinese and the Europeans developed an early awareness of and concern for the adequacy of forest resources. From this consciousness forestry evolved as a science.

Although the mature European economies pressed against their limited forest resource base, the dynamic economies of the New World seemed to possess unlimited forest resources. Isolated concerns were expressed occasionally, but a broad, general concern for the preservation and management of the forest did not arise in North America until the late 1800s, when anxiety over long-term timber availability led to the setting aside of national "timber reserves" in the United States.

This concern about future timber supply has continued to find expression both in the United States and worldwide. For example, the Forest Service of the U.S. Department of Agriculture (USDA) has undertaken assessments of domestic future timber availability at least six times in its history (Clawson, 1979), and assessment of the long-term timber supply situation in the United States is an ongoing responsibility of the Forest Service (see, for example, USDA, Forest Service, 1982). The Forest Service is not alone in its concern for and assessment of the adequacy of the U.S. timber resource; other agencies and organizations play important roles (see, for example, Irland, 1974).

In Europe estimates of future timber availability are made periodically by the Economic Commission for Europe/Food and Agriculture Organization (ECE/FAO) of the United Nations (ECE/FAO, 1986). In recent years other countries and international organizations have also developed projections of future timber availability. In addition to the work cited above, these efforts include those of the FAO Industry Working Party (1977), the Stanford Research Institute International (SRI) (1979), and the World Bank (1979). More recent efforts include that of the FAO in 1982 and that of the International Institute for Applied Systems Analysis (IIASA) (Kallio, Dykstra, and Binkley, 1987).

This book adds to the literature addressing the question of the long-term availability of timber and industrial wood. The principal tool of this study is the Timber Supply Model (TSM), which uses a control theory approach to investigate the availability and adequacy of the long-term economic timber supply from the major timber-producing regions of the world. Regional and world harvest levels, world market price, and investments in forest regeneration by region are projected for the fifty-year period 1985 to 2035. Various alternative scenarios are examined using simulations generated by the model.

THE CONVENTIONAL WISDOM

The conventional wisdom among most foresters today is that the forest resource is growing increasingly scarce. This view holds that demand for industrial wood will continue to expand in the future at a rate exceeding the expansion of the timber supply (see USDA, Forest Service, 1988a). In light of this situation, the real (inflation-adjusted) price of industrial wood will rise over time and with it the real price of stumpage.

To a large degree such a view is consistent with the experience of much of the past century and perhaps much longer. Although neither the United States nor the rest of the world has experienced "timber famines" such as have often been forecast, Potter and Christy (1962) found that forest resources—or at least some types of forest resources—have experienced increasing real prices in the United States going back to at least 1870. Other data document long-term real price rises in lumber beginning in 1800 (USDA, Forest Service, 1988b, p. 6). Because market prices indicate scarcity value (Barnett and Morse, 1963), rising real market prices indicate growing economic scarcity. Long-term data from other parts of the world are much more difficult to obtain, but those available exhibit a similar trend for other regions (see, for example, Sivonen, 1971).

Such findings are consistent with the general notion that rising population levels and economic activity will put increasing pressure on the earth's relatively fixed resource base, thereby causing increased economic scarcity that is reflected in real prices. For many years Forest Service projections reflected this view in its well-known "gap analysis," which measured a difference between projected levels of future consumption and projected future harvest. The basic projections, which assumed no real price changes, estimated the "shortfall" of output. Of course, the Forest Service recognized that an actual gap would never occur because in a market economy, such a situation would precipitate a price rise that would simultaneously choke off demand while inducing additional supply. Nevertheless, the gap analysis was useful as an indication of the direction of the pressure on prices and the nature of the market adjustments that would occur. The Forest Service has replaced the outmoded gap analysis with the Timber Assessment Market Model (TAMM) developed by Adams and Haynes (1980). This new approach also tends to generate projections that indicate a continuation of the long-term trend of rising real prices of timber (see, for example, Haynes, 1986).

A surprising finding of the work of Potter and Christy was that the forest resource (for which the proxy was lumber and pulpwood) was the only major natural resource to exhibit an important upward trend of real rising prices. For all other natural resources examined, the real price was essentially constant or declining. This finding was consistent with concurrent work by Barnett and Morse, who concluded: "Our empirical test does not support the hypothesis . . . that economic scarcity of natural resources . . . will increase over time in a growing economy" (p. 199).

These findings suggest that contrary to the popular perception, most natural resources are becoming economically more available over time. One apparent explanation for this contradiction of conventional wisdom for most natural resources is found in the role of technology, which extends the economic usefulness of physically limited resources. (This raises the question, addressed only partially in this volume, of why technology for dealing with the wood resource might have developed differently from technology used for other resources.)

Manthy (1978) reexamined and extended the Potter and Christy results up to the early 1970s. Expanding the indicators of the wood resource to include sawlogs, Manthy confirmed that the real prices of the forest resource, particularly sawlog prices, had indeed experienced a rising trend over the period examined and that this trend was at variance with the price trends of most other natural resources. However, Manthy also noted that not all wood resource prices were rising. Some wood prices—for example, pulpwood prices—did not exhibit a long-term upward trend like that experienced by sawlogs and lumber. In addition, Manthy also observed significant changes after 1950 in the upward trend of the prices of wood resources that had been experiencing increases. Manthy notes that "between 1950 and 1970 real forest product prices remained essentially stable. This stability was evident in each of the major forest product lines: sawlogs, veneer logs, and pulpwood" (p. 15).

A careful look at the post-1950 period from the vantage point of the mid-1980s does indeed suggest that the post-1950 period may be fundamentally different from the earlier period. Although the 1950s and 1960s exhibited essentially stable prices, the 1970s departed from this trend and exhibited very significant real price increases in industrial wood as well as most other natural resources. However, as will be demonstrated in chapter 3, the 1980s have seen a return to the relative stability in overall price levels of the 1950–1970 period; furthermore, the real prices of a variety of natural resource commodities, including industrial wood, have been declining. In the economic environment of the early 1980s, the real price of industrial wood first fell back to its pre-1970s level and subsequently remained relatively stable.

From this morass of prices and trends, several questions emerge. First, why was the long-term behavior of wood prices prior to 1950 so different from that of the prices of other natural resources? Second, in the post-1950 environment, has the real price behavior of industrial wood (especially sawlogs) changed, and is it beginning to conform with the more common trend of essentially stable prices of natural resources? Finally, how should we interpret the 1970s? Were the real price rises experienced by most natural resources (including wood) in the 1970s simply a nonrecurring perturbation generated by unique circumstances of the time—for example, the high level of inflation—or is another, more complicated explanation needed? This volume may not provide the definitive answer to all these questions, but we will draw some preliminary conclusions.

THE EXPANSION OF TIMBER SUPPLIES

All recent projections anticipate rising demand for and harvests of industrial wood. However, opinions differ on the rate at which demand is likely to increase. Furthermore, there are important differences in expectations as to the ability of wood-producing sectors to respond to increases in demand. The interaction between the growth in demand for industrial wood and the ability of the producing sector to provide the wood, and at what cost, ultimately determines the price of the wood resource. If the ability of the global system to expand its harvest of industrial wood is relatively limited, then modest increases in demand could precipitate rather large increases in real prices. If, however, the global wood-producing system can expand its harvests of industrial wood through time with relatively modest increases in (marginal) cost, then one might expect to observe relatively constant real wood prices over the long term.

This is not to say that the business cycle has been repealed. Wood prices would still be expected to rise during an upturn in the business cycle and decline to earlier levels over a downturn. The question addressed here is that of the long-term trend of real prices over several business cycles, not over one cycle.

In the face of rising long-term demand, there are several ways in which the wood-producing sector might respond. Historically, forest resources have been obtained largely by drawing down the naturally generated old growth forest. The Mediterranean Basin, for instance, was once occupied by large forests; the cedars of Lebanon were renowned in the ancient world for their size and utility in construction. Over the millennia the forest resource of the Mediterranean region was gradually mined as first the accessible low-lying forests were removed and later the less accessible forests (Thirgood, 1981). Simultaneously, many of these forestlands were converted to other land uses such as pasture and cropping.

The experience of other regions is similar in the sense that cutting progressed from more to less accessible sites and many forestlands were converted to other uses. For example, in the United States first the accessible forests of New England and the mid-Atlantic states were cut, then the Lake States and the pineries of the South, and finally the Pacific Coast forests, which became the target of logging to meet the industrial wood needs of the nation (Clawson, 1979). These historical examples demonstrate the process of utilizing existing accessible stands and then obtaining additional wood by moving to new forests that may be in more remote locations or more difficult terrain. In recent years Canada has experienced a similar phenomenon as the focus of logging has shifted from the Maritime provinces in the east to the accessible parts of the coastal forest of British Columbia and then to the less accessible forests of the interior of British Columbia and Alberta.

One element of this process that seems to have changed in recent decades is the apparent stabilization of the total forest area in the temperate regions of the world (Sedjo and Clawson, 1984). Forests are potentially renewable and expandable. The major impediments to reforestation are commonly continuing human disturbances that result in changes in land use away from forests and to another use, typically agricultural. At most times and places, forests regenerate naturally if the process is left undisturbed. Examples of massive natural regeneration abound. The denuded forests of New England and the Lake States have gradually been largely replaced by naturally regenerated second-growth forests. New England, which was less than 50 percent forested at the time of the Civil War, is now over 80 percent forested. In the U.S. South vigorous forests now grow where cotton and tobacco fields flourished for decades and even centuries. A similar process of return to forest has occurred in many places in Europe and the Nordic countries, and the total forest area of Europe has expanded over the past several decades as forests reclaim lands no longer useful in agriculture.

The renewability of forests is associated with the resurrection of industrial wood production in previously depleted regions. In the United States, for example, the center of the industrial wood production is shifting from the West back to the South. Similarly, the forests of the Lake States are once again providing industrial wood, this time for various types of composite panelboards. Thus, while society continues to shift its harvests to more inaccessible regions, it is also returning to areas that were previously harvested and have regenerated new forests that are now reaching maturity.

In the world context, massive additional stores of old growth forests remain in various places, including Siberia and the far eastern regions of the Soviet Union, northern Canada, and Alaska as well as tropical regions, especially the Amazon Basin and the central African forests. However, some of these regions (for example, Siberia) are highly inaccessible. Thus, economic logging and the associated transportation to major markets are costly and currently not justified by market prices that have existed in the past. In other of these regions and especially in the Amazon Basin and parts of the tropics, the lack of merchantability of the timber resource and/or extreme heterogeneity of species preclude economic exploitation. Although the area of tropical forest is clearly declining, this is largely the result of pressures for land use changes rather than pressures from commercial logging.

Not only are forests renewable, but humans can and do intervene in the process to make investments in reforestation, just as we make investments in cropping and other agricultural activities. The Chinese and Europeans have long been engaged in the process of forest management. Most European forests today are the product of forest management as well as natural processes. A similar situation obtains in many of the forests of eastern Asia, including Japan, Korea, and parts of China.

Although the U.S. South's earlier second forest resulted largely from natural regeneration, its current third forest is, importantly, the product of very substantial investments in artificial regeneration. In the Pacific Northwest and California, the old growth forests that are being logged are being replaced largely by artificially regenerated forests. Relatively new to the Western Hemisphere, artificial regeneration is now being practiced actively. In 1987 in the United States, 1.2 million hectares of forests were planted, including 1 million hectares in the South and hundreds of thousands of hectares in other regions.

In Latin America, hundreds of thousands of hectares of forest have been established artificially each year for the past twenty years. Worldwide in the 1980s, 14.5 million hectares of forestland or forest were being reforested or renewed annually (World Resources Institute, 1986). Much of that forest will probably never be available for industrial wood; however, much of the large-scale reforestation and management under way in traditional wood-producing regions is designed explicitly for the production of industrial wood.

In addition, massive investments in the establishment of industrial forest plantations are being made in tropical and semitropical regions that have not previously been important industrial wood producers. These plantations typically utilize a nonindigenous species. The largest of the "emerging" producers is Brazil, where hundreds of thousands of hectares of new industrial plantations have been established annually for the past fifteen or twenty years. Other significant establishments of industrial forest plantations are occurring in Chile, Argentina, Venezuela, New Zealand, Australia, South Africa, Spain, and Portugal.

TECHNOLOGY AND INNOVATION

If there is one feature that characterizes modern society, it is technological change. Although not generally viewed as a dynamic "high-tech" industry, the forest products industry is being profoundly affected by such change. Innovations permeate all facets of the industry, from the growing and harvesting of trees through the processing and into the development of new and improved products.

Because this process of change and innovation cannot but affect both underlying supply and demand for the resource and products of the industry, any attempt to assess the long-term adequacy of timber supplies must necessarily devote considerable attention to opportunities for technological change experienced by that industry. We have thus made special efforts to incorporate technological change into our analysis.

To a large degree the state of technology defines a resource, its boundaries in economic use, and its value. The history of the development of technology in the wood products industry is one of innovations that adapt the variable wood resource to the needs of the industry. Small logs have gradually replaced large logs as the feedstock for lumber and veneer mills as milling techniques have improved. Structural panelboards such as waferboard and oriented strand board made from hardwood chips of previously underutilized species are displacing plywood in many of its traditional markets. In pulp and paper production, techniques are gradually being developed that permit the utilization of wood fibers that were previously unusable. Southern pine fiber has become usable for newsprint; earlier, spruce was required. Gradually, innovations have allowed the substitution of short fiber for long fiber in the pulpmaking process. In the 1980s short fiber made up about 30 percent of the wood utilized in U.S. pulpmaking, whereas in the 1950s it made up only about 10 percent. Eucalyptus pulp, most of it from plantations in semitropical regions established since the 1960s, is now becoming the preferred pulp for many uses in Europe. Genetically superior seedlings give promise of even more rapidly growing trees, perhaps with other desirable characteristics. Such innovations can be characterized as "wood extending" in that they expand or extend the "wood basket" from which the economic supply is drawn.

In addition to allowing for the utilization of a wider variety of species, sizes, and qualities of wood, much technology in forest product processing has been of a "wood-saving" variety. Thus, although the demand for final products may expand at one rate, the demand for the underlying wood resource will expand at a lesser rate because of the continuing introduction of wood-saving technologies. Chapter 6 will discuss in greater detail the nature of various technological changes and estimates of their likely effects on industrial wood markets in the long term.

ENVIRONMENTAL CONSIDERATIONS

Tropical deforestation, acid rain (more correctly identified as air pollution), and global warming, subjects of growing international concern, could have major effects on forests. What effect might these phenomena have on this study's analysis and projections of long-term timber supply? In economists' jargon, the question is to what extent actual or possible environmental phenomena might change the biological production functions for industrial wood embodied in the model that underpins this study.

In attempting to address this question, we believe that it is important to note that an environmental impact on some portion of the world's forests does not necessarily translate into an impact on the world's industrial wood supply. Most of the world's forests are not important industrial wood suppliers. Environmental impacts will have significant direct effects on long-term timber supply only if they have significant impacts on forests and forest growth (biological production functions) that are a source of the industrial wood supply.

Tropical Deforestation and Fuelwood

Tropical deforestation is destroying about 7 million hectares per year of closed forest and 11 million hectares of both closed and open tropical forest.

An estimated 0.6 percent of total tropical forest is lost each year (FAO, 1982). The major cause of tropical deforestation is not commercial logging but rather pressure to convert forestlands to other uses, particularly agricultural (U.S. Department of State, 1980).

Although tropical deforestation is a serious environmental problem, current and even increased rates of deforestation are likely to have only a relatively modest impact on global timber supply, as tropical forests' current and future share of the total supply is small. These forests constitute roughly one-half of the world's total forested area and timber volume, but tropical timber makes up less than 10 percent of the world's industrial wood. A major reason for this is the heterogeneity of most tropical forest stands, which often renders their economic exploitation prohibitively expensive. Thus most tropical forests are not part of the world's industrial wood base, and even large changes in the area of tropical forests may not have significant effects on industrial wood production.

Furthermore, because this study formally incorporates into the TSM considerations of the effects of tropical deforestation on timber supply, continued deforestation is not an unanticipated external event that can upset the projections. (For example, the dominant supplier of tropical timber, the Asia-Pacific region, is specifically modeled in this study.) The modeling assumes that much of the existing tropical forest is now and will continue to be outside the timber base. Because many tropical forests are not economically available for industrial wood production, their destruction, while environmentally harmful, would have no effect on the model's projections and little effect on real-world harvest levels or price. The projections of the study suggest that tropical timber will continue to be harvested at about current levels or perhaps slightly lower levels over the next several decades. These levels are certainly attainable, given current or even significantly higher rates of tropical deforestation.

A related issue not explicitly addressed in this study is that of fuelwood in the Third World. As with the tropical deforestation issue, fuelwood availability is a serious local and regional problem (Sedjo and Clawson, 1984). However, the problem is largely confined to semiarid regions with few industrial forests. These regions are not now, nor are they anticipated to be, significant sources of industrial wood in the period under consideration. Hence they are not addressed directly in the analysis, nor will increased production of fuelwood in these regions have significant implications for industrial wood availability. Even in the one major timber-producing region in our study where fuelwood use is significant—the Asia-Pacific region—there is only limited direct competition between fuelwood and industrial wood as a result of locational considerations. In Indonesia, for example, although there are fuelwood problems on Java, the industrial forests are found literally hundreds and thousands of miles away on the islands of Borneo,

Sumatra, and New Guinea. Hence, to a very large degree the problems of industrial wood and fuelwood are independent.

Acid Rain

Acid rain or, more accurately, air pollution has been implicated in forest dieback in several regions of the world. The problem appears most serious in Europe, but it is also of concern in North America and elsewhere (Kairiukstis, Nilsson, and Straszak, 1987). For our purposes the question is the extent to which the forest dieback problem may affect the industrial wood supply and thereby invalidate the analysis and projections of this study. In the language of economics, the question again is the extent to which forest dieback modifies the underlying production function for industrial wood.

In the case of Europe, a significant portion of the dieback could occur in industrial forests. In this situation some of the assumptions of the study regarding the underlying European production function might no longer be valid. Although isolated local disruptions are unlikely to seriously disturb the global market or modify the conclusions of this analysis, a "large" dieback could have serious effects on global supplies. In these circumstances the assumptions of the model would be violated, and the projections and analysis based on the invalid assumptions would be of questionable relevance.

However, the assumptions of the TSM embodied in a regional production function can be modified to reflect changes associated with the dieback phenomenon and thereby simulate the long-term effects of a pollution-caused dieback. For example, by modifying yield relationships to capture the growth-inhibiting effects of pollution, Sedjo (1987) used the TSM to simulate the impact on the global market of large, long-term reductions in U.S. harvests as the result of a hypothesized major dieback in U.S. industrial forests. The simulations indicated that although some of the effects of U.S. harvest declines would be offset by increased harvests elsewhere, in the long term total worldwide harvests would decline and wood prices would rise. Economic losses would be relatively large in the region experiencing the dieback and smaller globally.

Global Warming

The aggregate effect of global warming on forests is poorly understood. The IIASA Global Trade Model (GTM) (Kallio and Dykstra, 1987) projected increased worldwide forest biomass and associated increased timber harvests as the boreal forests expanded in response to global warming; however, the efficacy of such projections is questionable. Aside from the question of the economic harvestability of forests located in the boreal region (few boreal forests are harvested in North America), forest ecologists anticipate that forest declines as well as expansions will occur in response to global warming (see, for example, Solomon, 1986). Although greenhouse experiments indicate that most plants, trees included, increase their short-term net photosynthesis with elevated levels of carbon dioxide (CO_2), the results of longer-term studies are more ambiguous (personal communication from B. R. Strain, cited in Sedjo and Solomon [1989]).

A recent attempt (Sedjo and Solomon, 1989) to assess systematically the net effect of global warming on forests by region used a Holdridge Life Zone approach together with a General Circulation Model for projecting climate changes. This approach generated estimates of forest biomass that were dramatically different from those of the IIASA model. These estimates indicated that both forest area and forest biomass would decline slightly in response to the higher temperatures accompanying a doubling of the earth's CO_2 level. The only clear conclusion that emerges is that there is no consensus among forest ecologists as to the effect of global warming on the total forested area of the globe.

Models such as the IIASA GTM or our own TSM have been or could be modified to attempt to capture the effects of global warming, but in light of the current state of knowledge, all such attempts must be suspect. This is not only because of the lack of consensus on the effects of warming on forests in the aggregate, but also because timber projection models in use are essentially steady-state models that assume stable production functions in which future forests and forest growth resemble those of the past or deviate in predicable ways—for example, yields will remain unchanged unless more inputs are used to generate more rapid growth. Global warming, if significant, would first generate a transition in the forests as they attempted to adapt to the new regime of temperature and climate by a change in species composition and so forth. Ultimately, the transition could so change the nature of regional forests and forest growth as to invalidate the parameters and relationships built into current timber supply models and therefore render existing projections of little use.

In light of the foregoing observations, when might the current projections of the TSM be of value in the context of a world experiencing global warming? If the warming occurs rapidly and becomes significant in the next decade, the projections of this study should be viewed with increasing skepticism. However, should the warming occur slowly over decades, the projections of this study, which are focused on the next two or three decades, might be only modestly affected. In the context of a major near-term global warming, the TSM may be useful if used as a simple timber harvest scheduling model to help project an economically rational drawdown of existing forests in anticipation of warming-induced change. It could also be used to simulate postwarming steady-state forest situations.

In summary, the effect of environmental changes on the projections of this study depends on the extent to which they impinge on the natural forest that is the source of industrial wood. This natural system is embodied in the model's production functions. Tropical deforestation is likely to have a negligible effect on the projections of this study because it will have little effect on the underlying production functions used in the model. Tropical timber is only a very minor portion of global industrial wood supply, most deforestation does not affect industrial supply sources, and a reasonable amount of tropical deforestation has already been assumed in the model. In contrast, pollution-caused forest dieback could generate significant deviations from the projections of the TSM by either (a) changing the relative shares of the several regions if a large dieback is confined to one region, or (b) generating a major change in global wood harvest levels and price if the dieback is large and widespread. Finally, a major global warming could so change the nature of the various forests as to render the underlying production functions irrelevant. In this case the current version of the model would be useless. However, should global warming occur slowly over several decades, warming may only modestly affect the production functions in the early period, and the TSM projections for the first two or three decades could be minimally affected.

THE PURPOSES OF THIS STUDY

This study has four purposes. The first is to present an improved model that develops a theoretical and methodological approach for addressing long-term timber supply problems that is more satisfactory than the approaches of current models. Such a new model must provide an explicit theory of the transition to a steady-state equilibrium. This should be done for a timber inventory possessing various age classes, allowing for the possibility of altering management intensity as a consequence of anticipated future prices. Hence the model should be capable of systematically examining long-term timber supply questions in a context that provides for the drawdown of existing stands of mature "old growth" timber as the system makes a transition from existing natural forests into managed, sustained-yield industrial forests.

None of the existing models does this in a conceptually satisfactory manner (Binkley, 1983). Most projection models, including the Forest Service TAMM and the IIASA GTM, simply shift a short-run, empirically estimated supply curve in response to changes in physical timber inventory (Binkley and Dykstra, 1987). Hyde (1980) has developed a model for looking at timber supply in the steady state in a world of forest management; however, his approach does not deal with questions of the drawdown of existing stands.

The TSM includes such considerations as the age of the inventory, opportunity costs of current and future harvests, harvest costs in different regions (including accessibility costs), and future growth and plantation possibilities to develop a long-run timber supply curve for a world in which both old growth and plantation forests coexist. By way of addressing the questions raised earlier, the model also addresses the issue of the economically optimal level and rate at which investments in future timber supplies should be made on forestlands of different productivity levels. Finally, the methodology incorporates changing technology into the analysis.

The second purpose of this study is to apply the model to real-world data to provide insights into the nature of the interactions that influence the long-term timber supply of industrial wood, and to examine a set of alternative futures to assess the sensitivity of a particular projection to changes in the underlying assumptions. That is, the effects of several different future situations are explored by examining various "alternative scenarios" with different but quite conceivable underlying assumptions. The scenarios examined include (a) alternative rates of growth in long-term worldwide demand, (b) alternative assumptions about the world's exchange rate structure, (c) alternative levels of newly established forest plantations in emerging producer regions, (d) an assumption of increased timber harvests in the Soviet Union, and finally (e) an assumption of a major fiscal "shock" on one major producer, the United States, in the form of higher taxes on forestry investment activities.

A third purpose of this study is to choose one of the set of feasible futures as representing a forecast of the most likely time paths of real prices and the volume of timber harvests. Over the past decade there has been a rather dramatic change in the perceptions of knowledgeable observers with regard to the long-term availability of timber resources both in the United States and worldwide. In the late 1970s the concern was over future availability; by the mid-1980s the concern had shifted to one of future markets and demand. In this context it is useful to have a forecast that integrates a broad array of relevant considerations and is based on articulated assumptions that can be modified, should conditions dictate. The projection used as our forecast is the "base-case" projection. It is designated as our "forecast" in that it represents our best judgment as to the future course of prices and harvests, given what we view as the most probable assumptions about future events, the data used in the model, and the conceptual consistency imposed on the projections by the model.

A forecast provides a perspective on the direction of long-term timber availability and particularly of prices. As such it is useful because in the process of making rational investment decisions in forestry, decision makers—both public and private—must have some notion of future timber prices and availability. These notions may be arrived at only informally and may enter the decision process implicitly, or they may take the form of quite formal and explicit forecasts. Because few single forecasts are likely to be consistently correct, an articulation of alternative views of the future, including those emanating from this study, can help provide decision makers with a more systematic basis for their decisions. This is particularly true in the context of the contemporary situation, which has seen marked changes in observers' assessments of the extent to which timber will be available in the future.

A final purpose of this study is to investigate the long-term availability of the timber resource in the spirit of the earlier studies of Potter and Christy (1962) and Barnett and Morse (1963). The focus here is on the timber resource, which is of particular interest because of the very different behavior of timber's historical real price trends when compared with those of most natural resources.

TOWARD A USEFUL MODEL

The Concept and the Problem

At the conceptual level the world's major industrial forest resources can be thought of as falling (not necessarily exclusively) into one of three categories of forest: old growth; secondary and managed forest; and planted, intensively managed forest plantations. Old growth forest refers to timber stands that are essentially virgin, having been relatively undisturbed by human activity. Because such standing timber has been allowed to accumulate over centuries, the wood volume in these stands is often very large. The major world regions with large old growth stocks of economic timber that are important current sources of timber include the Pacific Northwest of the United States and British Columbia in Canada; the forests of eastern Canada; the forests of the Soviet Union, both eastern and western; and the tropical hardwood forests of the East Indian Archipelago in Southeast Asia. Other regions that have vast inventories of old growth that are less important sources of industrial wood include Siberia, the Amazon forest, and the forests of central and western Africa.

Secondary and managed forests refer to major timber-producing regions where the forest is either second-growth forest, as the area had been logged or otherwise removed at some earlier time (but not so much earlier that the forest has acquired all the characteristics of an old growth forest), or a managed forest that has replaced an earlier indigenous forest. Forests of this type are found, for example, in the U.S. South, the Nordic countries, and in much of Europe and the Soviet Union.

The third group of forests are the planted and intensively managed plantation forests. These often consist of native tree species, as in the U.S. South and much of the Nordic region, although many plantation forests have introduced exotic (nonindigenous) species. Often these plantations are established on land that had not previously had important industrial forests. This is especially the case for plantation forests currently being established in the tropics and the Southern Hemisphere, particularly in Brazil, Chile, New Zealand, Australia, and South Africa, as well as in Spain and Portugal in Europe.

An examination of the forest groups defined above provides insights into an important phenomenon occurring today in forestry, one that the Timber Supply Model and this study are specifically designed to address. Gradually the world is experiencing a global transition from old growth forests to managed secondary and plantation stands. Early in human prehistory the pressures on the world's forests from wood needs were small compared to the large, naturally generated inventories and the forests' own ability to regenerate naturally. Human needs for the forest resource were met simply through collecting wood within the forest in a manner akin to the hunting and gathering mode utilized by early humans to meet their food needs.

Just as hunting and gathering gradually experienced a transition to livestock raising and cropping, so too modern forestry is experiencing a transition from the harvesting of naturally generated old growth stands to harvesting from forests that are managed actively and that are often the product of large investments in tree planting and tree growing.

This transition has several profound implications. First, whereas the economics of old growth forests are concerned primarily with the costs of harvest and transport, plantation forestry economics must be critically concerned with the biology and costs of planting, growing, *and* harvesting. Second, the decision about whether to make investments in regeneration can now be expanded to include the decision about where and when those investments might be made. Plantation forest investments now can and often are made in regions and areas that were not previously important industrial wood producers. In short, the location of the investments is not predetermined by the existence or lack of an earlier old growth forest. Third, with the cropping mode being applied to forestry comes the opportunity to develop and readily introduce a broad array of tree-growing technologies ranging from fertilizers to genetic improvements in the growing stock. Finally, the establishment of forest plantations at new geographic locations that were not previously important wood producers suggests a new structure of international trading patterns in forest resources and also processed wood products.

The problem, then, is to develop a model that captures the essence of the world timber system as it has been described here. Because in the real world there exist at any one time old growth regions, secondary forests, and plantations, the model must be able to deal simultaneously with all these forms of forest. In addition, the model must be able to project a region's changes as it makes its transition from an old growth to a secondary growth, managed, or plantation forest. Because the economics of timber harvesting remain important, the model must incorporate the relevant economic factors, including logging costs, access, and transportation for both old growth and other regions. Moreover, because the economics of timber growing will become increasingly important, the model needs to be able to introduce economically optimal levels of timber-growing investment into the system. Such a capability should reflect both existing and expected future prices as well as the biological responsiveness of the various sites. The model would need to be capable of projecting such a system for 50 to 100 years.

The Study's Approach

The Timber Supply Model is explicitly designed to examine and project the transition from old growth to secondary growth forest and to intensively managed plantations. The approach of this study is to place all the world's forest regions into one of two major categories, "responsive" or "nonresponsive," on the basis of economic, political, and data-availability considerations. Seven regions—the U.S. South, U.S. Pacific Northwest, western Canada (British Columbia), eastern Canada, Nordic Europe, Asia-Pacific, and the "emerging region"—are designated as responsive and are individually modeled and in many cases further subdivided; they are incorporated into the mathematical multiregional TSM. This model assumes economically optimizing behavior on the part of forest owners and managers in the responsive regions. Most of this study's analysis is focused on these seven regions, which currently account for about one-half of the world's industrial wood.

The formal model aggregates the supply of the seven regions and, together with world aggregate demand (adjusted for the effect of the other regions), simultaneously projects a unique time path of a unified world price. The world market is viewed as unified, hence consumption is not disaggregated by region. The model also solves simultaneously for the disaggregative intertemporal path of the harvests for each of the seven regions. For some regional subdivisions the harvest may be zero. In the process the transition from the existing forest to secondary growth forest with varying levels of management intensity is continued for each of the regions.

The rest of the world is defined as nonresponsive, in that either the regions are lacking substantial exploitable stands of timber resources or the economic/cultural system of the region is such that the region's timber production is unlikely to be responsive to economic forces and incentives. Because the production of the nonresponsive region is assumed to be determined largely by noneconomic considerations, the long-term growth of harvests from this region is determined exogenously, and the implications of alternative assumptions regarding this growth are examined for the world system through the TSM.

The TSM is designed specifically to track and respond to the changing conditions and considerations that apply to a "forest in transition" in a global context. A control theory approach is used for this task. The model first asks the question of how rapidly to draw down the existing forest inventory. The optimum rate of drawdown depends partly on the age distribution of the inventory. Old growth will typically be considered first for harvest. Unlike the typical linear programming model, the control theory approach does not require that the unrealistic drawdown of all the old growth be harvested immediately. Rather, it allows for a gradual though rapid drawdown, depending on events and conditions elsewhere in the world that affect the market. The model next determines the extent to which investments in regeneration for harvested areas should be made, based on economic criteria. Finally, the model determines the economically optimum rotation length for a region after it has harvested the old growth.

The Projections as Forecasts

In this study we treat the base-case scenario as our most likely scenario in a forecasting sense. This reflects our judgment that the set of assumptions embodied in this scenario is the most realistic.

An argument against using any of the projections as forecasts is found in the optimizing assumptions of the model. Although the projection technique assumes that all the harvests, management regimes, and so forth of the seven responsive regions are driven by economic optimization, this is clearly not the case for many subregions within the regions under examination. Also, the approach assumes that the nonresponsive regions ignore economic considerations, which again is clearly not the case for all subregions at all times. Hence it might be argued that because the assumptions are violated, the model cannot serve as a useful forecast.

However, the assumptions of this type of projection, or indeed of forecasts in general, are rarely if ever met perfectly. Over the long periods examined in this study (of up to fifty years), policies will invariably change in all the regions, both responsive and nonresponsive. Furthermore, although some subregions in the responsive regions may not be particularly responsive to market forces, the seven regions were chosen precisely because most of the industrial forestry activity in these designated regions is currently closely geared to forest products markets and is likely to remain so in the indefinite future. In addition, some of the lack of responsiveness of particular subregions is likely to be offset by responsiveness to economic incentives on the part of certain regions designated as nonresponsive. Even centrally planned economies are sometimes likely to respond to economic incentives that provide opportunities to earn foreign exchange. In short, even though the assumption of optimizing behavior on the part of all decision makers in the responsive regions is clearly an oversimplification, it is probably a good first approximation—we believe the best first approximation—of the forces that drive the world's timber-producing system over the long term. Hence, in our judgment aggregate projections based on that assumption as reflected in the base case are likely to be reasonable forecasts.

Given the foregoing considerations, we believe that the base-case intertemporal projections of harvests and price generated by the TSM also do provide a "most likely" aggregate forecast in a probabilistic sense. With respect to intertemporal harvest levels within the individual regions and subregions examined by the formalized model, however, we would expect the model's forecasting performance to be variable, depending on the extent to which conditions in individual regions and subregions conform to or deviate from the optimizing assumption. However, as the price projections depend on the relation of aggregate harvest to demand, the model should provide a good forecast of price as well as aggregate harvest. As with the projections, the forecast embodies the assumptions associated with the base case. Our judgment is that these assumptions are "highest probability" in the context of current circumstances, although invariably some of them will turn out to be in error to greater or lesser degrees.

THE ORGANIZATION OF THIS BOOK

Chapter 2 presents a broad overview of the general situation of the world's forests and a quick look at many individual regional forests. Particular attention is given to the forests on which this study focuses most of its attention.

Chapter 3 briefly discusses and analyzes the long-term behavior of price and consumption levels. In addition to providing a backdrop to the study, this chapter provides the basis for assumptions in the TSM regarding the growth of future demand.

Chapter 4 gives a nontechnical presentation of the TSM, which is the principal tool used in the analysis. Chapter 5 formally introduces demand and the growth of demand into the TSM. Chapter 6 examines the role of technology and some of its forms, and discusses its effect on the supply of industrial wood as well as on the growth of demand. Some of the currently emerging forest industry technologies are identified and their likely long-term impacts on supply and demand discussed.

Chapter 7 provides a technical discussion of the model and uses it to develop the elements of the elusive long-run timber supply curve. (This chapter may be passed over by the less technically proficient reader.) In addition, the serious technical reader will want to spend some time on the extension of the technical discussion of the model that appears in appendix O.

Chapter 8 presents the base-case projections together with several scenarios of alternative levels of growth in future demand. Chapter 9 expands the simulation approach to investigate various alternative scenarios involving differing assumptions about the exchange rate structure, the rate of establishment of new plantations, timber harvests in the Soviet Union, and the effect of tax changes on a major producing country, the United States. The various alternative scenarios simulate changes in basic assumptions and the occurrence of hypothetical but possible events and their effect on the long-term projections. The application of the model to alternative situations is limited only by available time and space. Its application to a contemporary policy problem, such as changes in the timber tax in the United States, demonstrates the broad utility of the TSM in examining the implications of a wide array of events (including policy changes) that might affect future timber availability.

Finally, chapter 10 develops many of the policy implications of the study and provides a concluding summary. The text is supported by appendixes A through O, which provide greater detail as to the specifics of the model and the data used.

REFERENCES

Adams, Davis L., and Richard Haynes. 1980. "The 1980 Softwood Timber Assessment Market Model: Structure, Projections, and Policy Simulations," *Forest Science Management* vol. 22 (suppl. to *Forest Science* vol. 26, no. 3).

Barnett, Harold J., and Chandler Morse. 1963. *Scarcity and Growth: The Economics of Resource Scarcity* (Baltimore, Md., The Johns Hopkins University Press for Resources for the Future).

Binkley, Clark S. 1983. "Optimization Models of Timber Supply," a comment on "Application of Optimal Control Theory to Estimate Long-Term Supply of Timber Resources," in Risto Seppala, Clark Row, and Anne Morgan, eds., *Forest Sector Models* (Berkhamsted, England, A. B. Academic Publishers) pp. 183–188.

Binkley, Clark S., and Dennis P. Dykstra. 1987. "Timber Supply," in M. Kallio, D. P. Dykstra, and C. S. Binkley, eds., *The Global Forest Sector: An Analytical Perspective* (New York, Wiley) pp. 508–533.

Clawson, Marion. 1979. "Forests in the Long Sweep of American History," *Science* vol. 204, no. 4398 (June) pp. 1168–1174.

Economic Commission for Europe/Food and Agriculture Organization of the United Nations (ECE/FAO). 1986. *European Timber Trends and Prospects to the Year 2000 and Beyond* (New York, United Nations).

Faustmann, Martin. 1849. "Berechnung des Werthes welchen Waldboden sowie nach nicht haubare Holzbestande fur die Waldwirshaft besitzen" ("On the Determination of the Value Which Forest Land and Immature Stands Possess for Forestry"), *Allgemeine Forst und Jagd-Zeitung* vol. 25, pp. 441–445.

Food and Agriculture Organization of the United Nations (FAO). 1977. *World Pulp and Paper Demand, Supply and Trade* (2 vols., Rome, FAO).

Food and Agriculture Organization of the United Nations (FAO). 1982. "World Forest Products: Demand and Supply 1990 and 2000," FAO Forestry Paper no. 28 (Rome, FAO).

Food and Agriculture Organization of the United Nations/United Nations Environmental Program (FAO/UNEP). 1982. "Tropical Forest Resources," FAO Forestry Paper no. 30 (Rome, FAO/UNEP).

Haynes, Richard W. 1986. "Future Supply and Demand for United States Timber," in Doris Robertson, ed., *Assessing Timberland Investment Opportunities, Proceedings 47346,* Forest Products Research Society (Madison, Wis., FPRS) pp. 27–32.

Hyde, William F. 1980. *Timber Supply, Land Allocation, and Economic Efficiency* (Baltimore, Md., The Johns Hopkins University Press for Resources for the Future).

Irland, Lloyd C. 1974. "Is Timber Scarce? The Economics of a Renewable Resource," bull. no. 83 (New Haven, Conn., Yale University, School of Forestry and Environmental Studies).

Kairiukstis, L., S. Nilsson, and A. Straszak, eds. 1987. *Forest Decline and Reproduction: Regional and Global Consequences.* Proceedings of a workshop held in Krakow, Poland, March 23–27, WP-87-75 (Laxenburg, Austria, International Institute for Applied Systems Analysis).

Kallio, Markku, and Dennis P. Dykstra. 1987. "Scenario Variations," in M. Kallio, D. P. Dykstra, and C. S. Binkley, eds., *The Global Forest Sector: An Analytical Perspective* (New York, Wiley) ch. 29.

Kallio, Markku, Dennis P. Dykstra, and Clark S. Binkley, eds. 1987. *The Global Forest Sector: An Analytical Perspective* (New York, Wiley).

Manthy, Robert S. 1978. *Natural Resource Commodities—A Century of Statistics* (Baltimore, Md., The Johns Hopkins University Press for Resources for the Future).

Potter, Neil, and Francis T. Christy, Jr. 1962. *Trends in Natural Resource Commodities* (Baltimore, Md., The Johns Hopkins University Press for Resources for the Future).

Sedjo, Roger A. 1987. "Pollution Related Forest Decline in the US and Possible Implications for Future Harvests," in L. Kairiskstis, S. Nilsson, and A. Straszak, eds., *Forest Decline and Reproduction: Regional and Global Consequences.* Proceedings of a workshop held in Krakow, Poland, March 23–27, WP-87-75 (Laxenburg, Austria, International Institute for Applied Systems Analysis) pp. 523–530.

Sedjo, Roger A., and Marion Clawson. 1984. "Global Forests," in J. Simon and H. Kahn, eds., *The Resourceful Earth* (New York, Basil Blackwell) pp. 128–170.

Sedjo, Roger A., and Allen M. Solomon. 1989. "Climate and Forests," in N. Rosenberg, W. Easterling, Pierre R. Crosson, and Joel Darmstadter, eds., *Green-*

house Warming: Abatement and Adaption (Washington, D.C., Resources for the Future).

Sivonen, S. 1971. "Havusahapuun Rantohinnan paa Sunntrainen Reyitya Suomessa Vuosina 1920–67," Report of the Committee on the Costs of Forest Planting and Seeding, Annex 5, Helsinki, Finland. *Folia Forestalia* vol. 109, pp. 116–120.

Solomon, Allen M. 1986. "Transient Response of Forest to CO_2-Induced Climate Change: Simulation Modeling Experiments in Eastern North America," *Oecologia* vol. 68 (Spring) pp. 567–579.

SRI International. 1979. "The Outlook for the World's Forest Products Industries," private report (Palo Alto, Calif., SRI International).

Thirgood, J. V. 1981. *Man and the Mediterranean Forest—A History of Resource Depletion* (London and New York, Academic Press).

U.S. Department of Agriculture (USDA), Forest Service. 1982. "An Analysis of the Timber Situation in the United States 1952–2030," Forest Resource Report no. 23 (Washington, D.C., USDA).

U.S. Department of Agriculture (USDA), Forest Service. 1988a. "The South's Fourth Forest: Alternatives for the Future." Forest Service Report no. 24, p. 512 (Washington, D.C., USDA).

U.S. Department of Agriculture (USDA), Forest Service. 1988b. "U.S. Timber Production, Trade, Consumption, and Price Statistics 1950–1986," miscellaneous publication no. 1453 (Washington, D.C., USDA).

U.S. Department of State. 1980. *The World's Tropical Forests: A Policy, Strategy, and Program for the United States.* Report to the president by a U.S. interagency task force on tropical forests (Washington, D.C., U.S. Government Printing Office).

World Bank. 1979. EIS Paper no. 98, prepared by James Gammie (London, IBRD).

World Resources Institute. 1986. *World Resources 1986* (New York, Basic Books).

2

World Forest Resources and Production

This chapter provides an overview of the world's forest resources and industrial wood production. Particular attention is given to the seven "responsive" regions that are currently or are projected to be major producers of the world's industrial wood. In addition, this chapter briefly examines the major producing regions that are not modeled in the study (that is, the "nonresponsive" regions), focusing attention on the large industrial wood-producing areas.

The world's uses of wood can be divided into two major categories—industrial and fuel—with total worldwide production divided roughly equally between them. In the industrial world most wood used is for industrial products, including solidwood products such as lumber and wood panels as well as fiber for pulp and paper products. In the developing world, just the reverse is the case: most of the wood is used for fuel, either directly or when processed into fuels such as charcoal.

Table 2-1 gives the world's production (harvest) of roundwood in 1980 by major region. Although each region has significant production, about 75 percent of the world's production is in the temperate regions of North America, Europe, and the Soviet Union.

The mix of roundwood uses differs dramatically across countries and regions. In North America, for example, about 95 percent of harvested wood is used for industrial purposes. By contrast, in Africa this amount drops to 12 percent. There are exceptions, however; for example, the high-quality, accessible tropical hardwood resources of the Asia-Pacific region are being harvested largely for export markets.

There is a poor correlation between industrial wood production and the simple existence of large forested areas. Although forest resources are obviously necessary for industrial wood production, wood volumes are not a sufficient condition for production. For a variety of reasons, many of the most heavily forested regions are not important industrial wood producers. In the case of the Amazon, for example, the heterogeneity of the wood resource limits its commercial uses. Hence, although about one-half of the world's natural forest is found in tropic and semitropic regions, less than 10 percent of the world's industrial wood comes from these tropical forests. Similarly, in Soviet Siberia and northern Canada, for example, physical wood resources alone are not a sufficient condition for industrial wood production. The cost of access limits the forests' commercial value, and large volumes of potentially usable industrial wood are left undisturbed.

The focus of this study is industrial wood production. The regions modeled produced roughly one-half of the world's industrial wood in the 1980s. They are dominated by four North American regions and also include the Nordic region of Europe, the tropical hardwood forests of the Asia-Pacific region, and the emerging forest plantations of the Southern Hemisphere and the Iberian Peninsula. The forests of other regions that are important industrial wood producers are also examined in this chapter. These include forests in the Soviet Union and non-Nordic Europe as well as other forests spread across the globe.

The first sections of this chapter present the aggregate state of and trends in the world's forests. Subsequently, the forest situation in the major forested regions of the world is examined in somewhat greater detail. The major purpose of this chapter is to give the reader a sense of (1) the relative current

Table 2-1. Roundwood Production by Region, 1980

	Total roundwood production		Industrial roundwood production	
Region	(thousands of m^3)	(percentage)	(thousands of m^3)	(percentage)
Africa	433,851	14.4	50,430	3.7
North America (United States and Canada)	483,631	16.0	463,958	33.3
Central America	47,208	1.6	10,958	0.8
South America	315,202	10.4	65,922	4.7
Asia	1,017,100	33.7	206,696	14.8
Europe	334,346	11.1	291,321	20.9
Oceania	32,968	1.1	25,986	1.9
Soviet Union	356,000	11.8	278,200	20.0
World	3,020,306	100.0	1,393,471	100.0

Source: Data from Food and Agriculture Organization of the United Nations (FAO), *Yearbook of Forest Products 1969–1980* (Rome, FAO, 1980).

volumes of industrial wood in the major regions, (2) the biological growth rates likely in these regions, and (3) the anticipated direction of changes in forest volumes over time. These, together with later discussions in this text, should illuminate the potential role of each of the regions in generating long-term timber supplies. Because the quality of the available data varies considerably among regions, the discussions are not always symmetrical or supplied with the same supporting data.

AN OVERVIEW OF THE WORLD'S FORESTS

Forests cover about 31 percent of the world's land surface. Today about 2.7 billion hectares (ha) are covered by closed forest (table 2-2), that is, forest that has a substantially complete cover of trees over the whole surface of land. The remainder consists of less densely wooded open forests (1.2 billion ha). In addition, when forest regrowth on fallowed cropland covering an additional 400 million ha, and natural shrublands and degraded forests in developing countries covering another 675 million ha, are included, the total area of "woody vegetation" is about 5.2 billion ha, or about 40 percent of the world's land area (World Resources Institute, 1986, p. 62).

About 70 percent of the world's wood used for industrial purposes is coniferous (softwood), and over 90 percent of the coniferous forest is located

Table 2-2. World Forested Area by Region, Circa 1985

	Forestland	Closed forest	Open woodland	Total land area	Closed forest	Total forest
	(millions of hectares)				(percentage of land area)	
North America	734	459	275	1,829	25	40
Central America	65	60	2	272	22	24
South America	730	530	150	1,760	30	41
Africa	800	190	570	2,970	6	27
Europe	160	148	12	472	31	34
Soviet Union	930	792	138	2,240	35	42
Asia	530	400	60	2,700	15	20
Pacific area	190	80	105	842	10	23
World	4,139	2,659	1,200	13,105	20	31

Note: Forestland is not always the sum of closed forest plus open woodland, as it includes scrub and brushland areas that are neither forest nor open woodland as well as deforested areas where forest regeneration is not taking place. In computation of total land area, Antarctica, Greenland, and Svalbard are not included; 19 percent of all Arctic regions are included.

Source: Data from Economic Commission for Europe/Food and Agriculture Organization of the United Nations (ECE/FAO), *The Forest Resources of the ECE Region* (Geneva, ECE/FAO, 1985).

in the northern temperate regions of North America, Europe, and the Soviet Union. These regions, not surprisingly, are the dominant producers of industrial wood. Worldwide, however, about 60 percent of the forestland area is nonconiferous (hardwood), and these forests are more evenly spread around the globe. Although South America and Asia account for roughly 50 percent of the total nonconiferous forest, sizable forest areas are also found in North America, the Soviet Union, Africa, and elsewhere (table 2-3).

Increasing attention has been given to reforestation activities around the world over the past decade. Table 2-4 presents an estimate by the Food and Agriculture Organization (FAO) of the United Nations of the amount of plantation forest by major region in the mid-1970s. At that time the portion of the world's total closed forests attributable to plantations was only about 90 million ha, or 3.4 percent of the world's total closed forest. The area of forest plantations and reforestation has increased substantially, however, since the World Resources Institute (1986) estimated that 14.5 million ha were being reforested or renewed annually by the beginning of the 1980s. Large areas are planted each year throughout the world. In traditional industrial wood-producing countries, the Institute's (1986) estimates of reforestation efforts in recent years include Canadian plantings of 720,000 ha per year and

Table 2-3. Land Area of World Forest Resources (Closed Forests) by Region and Type
(land area in millions of hectares)

	Coniferous forests		Broadleaf forests		Combined coniferous and broadleaf forests	
Region	Land area	Percentage	Land area	Percentage	Land area	Percentage
North America	400	30.5	230	13.4	630	20.8
Central America	20	1.5	40	2.3	60	2.0
South America	10	0.8	550	32.0	560	18.5
Africa	2	0.2	188	10.9	190	6.3
Europe	107	8.2	74	4.3	181	6.0
Soviet Union	697	53.1	233	13.6	930	30.7
Asia	65	5.0	335	19.5	400	13.2
Oceania	11	0.8	69	4.0	80	2.6
Total world	1,312	100.0	1,719	100.0	3,031	100.0

Note: The totals for combined coniferous and broadleaf forests do not always add because no breakdowns have been given for areas in Europe and the Soviet Union excluded by law from exploitation.

Source: For Europe and the Soviet Union, data from Economic Commission for Europe/Food and Agriculture Organization of the United Nations (ECE/FAO), *The Forest Resources of the ECE Region* (Geneva, ECE/FAO, 1985). For the remainder, data from Reider Persson, *World Forest Resources: Review of the World's Forest Resources in the Early 1970s,* no. 17 (Stockholm, Royal College of Forestry, 1974).

Table 2-4. Plantation Forest by Region, Circa 1975

Economic class and region	Hectares (million)
Developed	
North America	11
Western Europe	13
Oceania	1
Other	10
Total	35
Developing	
Africa	2
Latin America	3
Asia	3
Total	8
Centrally planned	
Europe and the Soviet Union	17
Asia	30
Total	47
Total world	90

Source: Food and Agriculture Organization of the United Nations (FAO), "Development and Investment in the Forestry Sector," FO:COFC-78/2 (Rome, FAO, 1978). Reprinted with permission.

Japanese plantings of 240,000 ha per year; in addition, the Institute cites Soviet reports of reforestation of 4.5 million ha in the Soviet Union in 1980. U.S. Department of Agriculture statistics (USDA, 1988b) indicate that 1.2 million ha were planted in the United States in 1987. Also, very substantial forest plantations are being established in such countries as Brazil (up to 450,000 ha per year in the late 1970s), Chile, New Zealand, and elsewhere. Finally, China has reportedly planted tens of millions of hectares of trees over the last two decades.

Forests in the Temperate Regions

About one-half of the world's more than 4 billion hectares of forestland are located in the temperate regions, and these are predominantly in the Northern Hemisphere. These forests provide the vast majority of the world's industrial wood. Of the seven regions modeled in our study, five are found in the temperate region of the Northern Hemisphere. In addition, most of the industrial wood of the nonmodeled regions also comes from Northern Hemisphere temperate forests, which include those of the Soviet Union, Europe, and Japan.

The Southern Hemisphere's temperate forests are much less important in terms of standard measures such as forestland area, inventory volumes, and timber production; however, recent industrial forest plantation activity suggests the possibility of large volumes of future production from this region. It is largely the temperate regions of the Southern Hemisphere—for example, Chile, New Zealand, South Africa, and parts of Brazil—that are modeled as the so-called emerging region and give promise of being major industrial wood producers in coming decades.

Despite the fact that it provides most of the world's industrial wood, the land area reported in the temperate forests actually increased about 2 percent in the post-World War II period (Sedjo and Clawson, 1984). A recent study (World Resources Institute, 1986, p. 65) shows that almost all the countries of Europe experienced increases in their forest areas from 1954 to 1984; the total European forest area experienced an increase of over 15 percent during that period. Other regions experiencing the largest increases in forest area include China, the Soviet Union, South Korea, and Oceania. Growth in these areas more than offset modest declines in forest area in North America. Those regions which the data indicate are expanding their land area in forests are often regions experiencing active reforestation efforts. In other cases, however, the expansion of forest area represents a resurgence of forest that often fills the void left by declining agriculture.

Tropical Forests

Lying on each side of the equator around the world is an immense area of tropical forests that constitute almost one-half of the world's total forestland. Although there are commercial timber harvests in many areas of the tropics—most notably the Asia-Pacific region, some countries of Southeast Asia, and parts of West Africa—tropical forests are not nearly as important a source of commercial timber as their area and forest volumes might suggest. This is due largely to the great heterogeneity of species in the forest, which, coupled with the noncommercial nature of many of the species, severely limits their exploitation. Of the tropical forests, only those of the Asia-Pacific region are modeled in this study. This region accounts for less than 5 percent of the world's total industrial wood production; however, it is by far the most important single region providing tropical hardwood timbers.

The tropical forests consist of both moist forests and dry forests. A third of the tropical moist forests are in Brazil—the Amazon Basin—and another quarter are in other Latin American countries. Other major areas are in west and central Africa, Asia, and the islands of the East Indies. The tropical dry forests are located largely in Africa and parts of Latin America, with smaller areas in Asia. There are of course some biological differences among these areas. Most of these tropical moist forests are characterized by extremely high volumes of vegetation per unit area and by great diversity in the vegetative cover in each area.

The tropical dry forests are even less important as a source of industrial wood than their physical presence would suggest. However, in many regions these forests are an important source of fuelwood. The problems of deforestation associated with excessive fuelwood collection largely occur in areas of tropical dry forest such as the Sahel in Africa and parts of India and Nepal.

Industrial Plantation Forests

Just as humans gradually made a transition in food production from foraging and hunting to cropping and livestock raising, so society is gradually moving away from a dependency on naturally created forests for industrial wood needs toward plantation and managed forests. Industrial plantation forests, that is, artificially generated man-made forests, can be established subsequent to a harvest of a natural forest to increase the rate of regeneration, ensure the establishment of desired species, or provide superior trees for more rapid growth and better form. In addition, industrial plantations can be established as a crop on depleted agricultural land or, in some cases, to create a forest where nature had never placed one.

In the regions that are modeled in this study, several of the forests are gradually experiencing a transition to plantation forests as the old growth is drawn down and replaced by man-made plantation forests. This is occurring in the United States, Nordic Europe, and elsewhere. In addition, one region—the emerging region—depends on the establishment of plantations of exotic, nonindigenous species. This region plays an important role in the Timber Supply Model, as it provides a new and growing source of industrial wood. In the emerging region particularly, large areas have been and are continuing to be established as plantation forests. The economics for many of these plantations appear favorable (Sedjo, 1983).

The potential impact of plantation forests on industrial wood supply can be seen from the recent experience of Latin America. Although less than 1 percent of the forests of Latin America are industrial plantations, about one-third of the region's industrial wood output comes from industrial forest plantations. Furthermore, the total area in industrial forest plantation in Latin America is projected to increase 300 percent between 1979 and the year 2000. By the year 2000 it is expected that more than half of the greatly expanded industrial wood production of Latin America will come from plantation forests (Inter-American Development Bank, 1982, p. 17).

A similar trend toward plantation forests exists in the United States. Preliminary projections of the Forest Service indicate that by the year 2000, pine timber removals from plantations will exceed the removals from natural pine stands in the South (USDA, Forest Service, 1988a). Table 2-4 suggests that the creation of U.S. plantation forests is quite widespread.

The foregoing data not only attest to the potential of forest plantations to replace much of the natural forests as the principal source of industrial wood, but also suggest the extent to which relatively small areas of highly productive forest plantations can substitute for larger areas of natural forests as producers of society's industrial wood needs. Therefore, simple comparisons of deforested areas and areas in forest plantations must be made with care.

It is only in very recent times that forest plantations have been a significant factor affecting the world's forests. Although some conscious tree planting took place in parts of China as far back as the fourth century B.C., most reforestation has occurred through natural regeneration. However, the incidence of artificial regeneration has increased dramatically since World War II and particularly after 1960.

Forest plantations are the result of conscious management. Commonly, the plantations of the temperate Northern Hemisphere are established on land that has recently been logged. Often the plantation is in the same species as the harvest: for example, Douglas fir in the Pacific Northwest and Scots pine in the Nordic countries. However, other exotic or nonindigenous species may be introduced: for example, lodgepole pine in parts of northern Europe. In many cases in the Northern Hemisphere, the plantation is established on lands that have not recently been forested. This is the case, for example, in parts of the U.S. South where farmlands have been converted to forests. In China and Great Britain, plantations are sometimes being established on lands that have not been forested for centuries.

In recent years increased attention has been given to the establishment of forest plantations in the tropics, subtropics, and the Southern Hemisphere. Although it is commonly believed that massive areas of plantation are being established in the tropics, the data indicate that only a small portion of forest plantations are situated in the tropics. Rather, the majority of the plantation lands are in the subtropics and typically are established on lands that have not been forested in recent years. In some cases the lands have never been in forest, for example, the Orinoco River Basin in Venezuela. Today major industrial forest plantation activities are under way in Brazil, Chile, Venezuela, South Africa, India, the Philippines, Australia, New Zealand, Spain, Portugal, and a host of other tropical, subtropical, and Southern Hemisphere countries. The dominant species being planted are exotics, predominantly tropical pines, North American pines, and eucalyptus. Brazil alone established an average of over 250,000 ha per year of exotic forest plantations during the 1970s, and Spain and Portugal are becoming important forest resource producers through recently established eucalyptus plantations. Projections by Lanly and Clement (1979) for the year 2000 indicate that the industrial plantations in the tropics and subtropics will cover over 21 million ha, or three times the area in forest plantations in the mid-1970s.

Forest plantations in the semitropics and nontraditional areas offer great promise; however, a cautionary note is in order. Because experience is limited, there are some concerns over the possible ecologically serious problems involving insects, disease, and so forth. Fortunately, actual experience has been encouraging thus far; there have been few really serious problems of disease or insect infestation.

MAJOR INDUSTRIAL WOOD-PRODUCING REGIONS

Europe

The European Forest

Europe is one region where the forestland area and growing stock have been expanding. This has certainly been true since World War II and probably since the mid-nineteenth century. The development of vegetative types in the European region is related not only to such evident factors as the diversity of climate and soil conditions but also to the prolonged influence of humans as a consequence of a high density of population over several centuries.

In the north are the boreal forests, which cover large areas in the Nordic countries of Finland, Sweden, and Norway. These forests, identified as the Nordic region, are formally introduced as one of the seven regions in our TSM. The remainder of Europe—with the exception of the exotic forest plantations of Spain and Portugal, which are incorporated into the emerging region examined in the model—is treated as a nonresponsive region even though it is heavily forested and an important industrial wood-producing region. Hence it is not formally introduced and examined in the context of the modeling effort.

Pine and spruce predominate in the species composition of the boreal forests of northern Europe. The subalpine forests dominated by evergreen conifers extend far south, even into the Mediterranean peninsulas, as air temperature decreases with increasing altitudes. Pine, spruce, larch, and fir constitute most of the subalpine forests of western and central Europe extending southward into southern Europe.

The ecotonal mixed forests constitute a transition between the previous types of coniferous forests and the nonconiferous deciduous communities of the humid regions. The main species are pine, spruce, silver fir, oak, birch, and beech.

Finally, the evergreen mixed forests, which were once widely spread over the Mediterranean area, have gradually been modified by intense human activity and are now confined to relics or to degraded forms. The original woodland vegetation now appears mainly as scattered clumps of holm oak, cork oak, and pine.

Europe's Industrial Wood Resource

Table 2-5 indicates that of the total land area in Europe (472 million ha), 160 million ha, or about one-third, are classified as forest and other wooded land. The Nordic countries have the highest proportion of land covered by

Table 2-5. Major Land Use Categories by Region in Europe, Late 1970s to Early 1980s
(land area in millions of hectares)

Region	Total land (excluding water)	Forest and other wooded land[a] (percentage of total land)	Exploitable forest (percentage of forest and other wooded land)	Unexploitable forest	Other wooded land	Nonforest land	
						Agricultural	Other
Nordic countries	112.4	59.9 (53%)	48.3 (81%)	3.6	8.0	9.8	42.7
Western Europe	234.2	60.6 (26%)	42.6 (70%)	7.5	10.5	131.0	42.6
Eastern Europe[b]	125.4	39.8 (32%)	35.6 (89%)	1.1	3.1	73.6	12.0
Total[c]	472.0	160.3 (34%)	126.5 (79%)	12.2	21.6	214.4	97.3

Source: Economic Commission for Europe/Food and Agriculture Organization of the United Nations (ECE/FAO), *The Forest Resources of the ECE Region* (Geneva, ECE/FAO, 1985). Reprinted with permission.

[a] Other wooded land is defined as having a tree cover of less than 20 percent.

[b] Including Albania and Yugoslavia.

[c] Excluding Cyprus, Israel, and Turkey.

forests and other woods (53 percent), followed by eastern Europe with 32 percent and western Europe with 26 percent.

Forest exploitable for industrial wood production covers some 126 million ha in Europe, or 79 percent of the total forest and other wooded land. In all three regions the proportion of exploitable forest in the total is roughly the same.

Unexploitable forest covers about 12 million ha in Europe, or 8 percent of the total of forest and other wooded land. The reasons for classifying forests as unexploitable may include either legal restrictions on commercial cutting because of protective and other non-wood-producing functions or economic or physical inaccessibility.

For many centuries the forests of Europe were cleared to make space for an expanding agricultural economy. Toward the end of this phase of forest contraction, large quantities of wood were required to meet the demand for industrial and domestic fuel as well as to provide wood for construction and shipbuilding. The turning point, that is, the point at which the area of forest began to increase again in Europe as a whole, probably came during the nineteenth century, although it varied from country to country according to the pace of industrialization and substitution of other forms of fuel for wood (ECE/FAO, 1976).

The Nordic Countries

Within Europe the Nordic countries constitute a very important producer and exporter of industrial wood, feeding the major markets in Great Britain and on the Continent. The Nordic countries—Sweden, Finland, and Norway—have been treated as a responsive region within the TSM; therefore, they have been modeled separately.

Being northern climatic regions, the Nordic countries experience relatively modest forest growth. However, their forested area is substantial, and total production exceeds 5 percent of total world production. In 1980 the Nordic countries had 38 percent of Europe's exploitable forest, 29 percent of the growing stock, and 31 percent of the growth increment. Growing stock on exploitable lands in the Nordic countries amounted to 4,407 million cubic meters (m^3) with a net annual incremental growth of 146 million m^3. This corresponds to an average standing volume of 91 m^3/ha with a net annual incremental growth of 3.0 m^3/ha, both well below the European average. Despite this relatively low level of annual increment per hectare, the Nordic countries have an average (3.3 percent) rate of increment in relation to growing stock. Coniferous species represent approximately 84 percent of growing stock (ECE/FAO, 1985).

Western Europe

Western Europe held over one-third of Europe's exploitable forestland and a similar proportion of the growing stock in 1970. France and Germany alone account for almost one-half of the region's forest inventory. There is a high proportion of conifers in Germany, where growing stock per hectare (145 m^3) and the increment per hectare (4.8 m^3) are well above the average. Conversely, broadleaved species account for nearly two-thirds of the growing stock in France, where the standing volume and increment per unit area are much closer to the average. Switzerland has by far the highest volume per unit area of exploitable forest anywhere in Europe: 392 m^3/ha. This amount reflects in part the Swiss forest structure; Swiss forests are typically highly dense and have trees of fairly large diameter with a high volume per hectare. Austria also has a high growing stock per hectare.

Some countries, such as Ireland, have a high proportion of young plantations of quick-growing species; those countries thus have unusually high annual increments. Italy has a fairly extended area of exploitable forest but a low standing volume and low net annual increment. Portugal and Spain have relatively high rates of growth, in part because of the importance of fast-growing exotic plantations and also because of favorable climatic conditions.

Eastern Europe

The proportions of growing stock and increment in this region differ from those of other exploitable forests of Europe. In about 1980, eastern Europe accounted for 28 percent of the exploitable forest area, 36 percent of Europe's growing stock, and 28 percent of the increment. These percentages indicate a volume of growing stock of 155 m^3/ha, well above the European average. The net annual increment is at 133 million m^3, which is equivalent to 3.7 m^3/ha, or 2.4 percent of the growing stock volume—the highest of the three regions.

In terms of growing stock, eastern Europe also has a balanced mix of hardwoods and softwoods, with coniferous species accounting for a little over half of the growing stock and annual increment. However, the composition of the standing volume and the share of conifers and broadleaved species varies substantially among the countries within the region. Czechoslovakia, the German Democratic Republic, and Poland are the predominantly coniferous countries in contrast to Bulgaria, Hungary, Yugoslavia, and Romania, where the larger part of the growing stock is broadleaved (ECE/FAO, 1985).

Soviet Union

The Soviet Union is the world's leading nation in terms of the extent of forested area and forest resources within its borders, and after the United States it has the second-highest volume of industrial wood production. Because its economy is centrally planned, the Soviet Union was not included among the responsive countries examined in this study. Nevertheless, the large volumes of existing timber and the highly political nature of production decisions offer the Soviet Union the potential to significantly increase its harvests. Although Sutton (1975, p. 110) doubts the ability of the Soviet Union to significantly increase its industrial wood production, Holowacz (1985, p. 367) states that this nation can "almost double its annual cut without undermining the productive capacity of the existing forests." Large increases in harvests undoubtedly would affect world markets and thus influence the production levels of other producing regions. One of the scenarios discussed later in this book examines such a situation.

In 1978 the Soviet Union accounted for 21 percent of the world's forested area and for more than 25 percent of the world's growing stock (United Nations Industrial Development Organization, 1983). The softwood growing stock in Soviet forests makes up one-half of the world's total. About one-third of the world's volume of temperate hardwood growing stock is also among the Soviet Union's forest assets (Persson, 1974).

Although detailed data on Soviet forests are sometimes difficult to obtain, Sutton's breakdown of the exploitable forest area in 1966 by major region and type is presented in table 2-6. Estimates of the total volume of wood from potentially exploitable forests were 47 billion m^3 in 1964 and 48 billion m^3 in 1966. A subsequent estimate in 1969 of the total volume in then currently accessible reserves was 36.5 billion m^3, which indicates the difference between what was accessible and what was potentially exploitable around the mid- to late 1960s.

Holowacz (1985) provides updated estimates of exploited forest area, the levels of regeneration, and the total area of forest in the Soviet Union. The data indicate that the exploited forest area was 385.3 million ha in the early 1980s. In addition, the total Soviet forest area has increased over the past

Table 2-6. Area of Exploitable Forests in the Soviet Union, 1966
(millions of hectares)

	Forest area			
Forest type	Total USSR	In European region	In Siberia and the Far East	Elsewhere
Coniferous species, excluding larch	147.1	73.5	72.6	1.0
Larch	84.6	—	84.6	—
Total coniferous	231.7	73.5	157.2	1.0
Total nonconiferous	96.7	44.3	47.4	5.0
TOTAL EXPLOITABLE	328.4	117.8	204.6	6.0

Source: W. R. J. Sutton, "Forest Resources of the USSR: Their Exploitation and Potential," *Commonwealth Forestry Review* vol. 54, no. 160 (June 1975), p. 112. Reprinted with permission.

Table 2-7. Increases in Soviet Forest Area, 1961–1984

Year	Area of forest land (millions of hectares)	Volume of growing stock (millions of m^3)
1961	738	80,200
1973	769	81,900
1978	792	84,200
1984	811	85,900

Source: Adapted by permission of the publisher, the Canadian Institute of Forestry, from table 2 in J. Holowacz, "Forests of the USSR," *Forestry Chronicle* vol. 61, no. 5 (October 1985), pp. 366–373.

three decades, resulting in an increasing trend in forestland and growing stock between 1961 and 1984 (see table 2-7). The expansion in forest area and growing stock reflect the low level of cut relative to growth. Currently only about 35 percent of the total net annual growth is utilized, and even adjusting for accessibility, this is well below technically and biologically sustainable levels of cut. An additional reason for the increase in area and volume is that "the USSR regenerates annually an area larger than it clear-cuts," with regeneration during the 1981–1985 period estimated at 10.7 million ha. Although forests in some parts of Soviet Europe are experiencing severe local pressures, the forests east of the Ural Mountains have experienced little serious exploitation.

In contrast to Sutton, Holowacz finds the Soviet Union in a position of being able to increase dramatically its annual cut. He notes, however, that forest production does not appear to be a high priority and that Soviet forest-based industries are primarily domestically oriented.

The Soviet forest resource is disproportionately distributed in relation to the population and location of wood-processing facilities. This allows the overall resource situation to be favorable even as some parts of Soviet Europe are experiencing severe local pressures. The area northeast of a line extending from the city of Leningrad to the junction of the Soviet, Chinese, and Mongolian borders holds over 85 percent of the forested area along with only about 14 percent of the country's population. The area south of the line supports the remaining 15 percent of the forested area and about 86 percent of the population. The southern region relies heavily on wood resources brought in from the north and the east, although there is a trend toward relocating forest-based industries in the densely forested region (Holowacz, 1974, 1985).

Siberia and the Far East dominate the forest picture, constituting 76 percent or 515 million ha of the forested land and 79 percent or 53 billion m^3

of the wood reserve. They also have the highest proportion of volume in mature stands but the poorest growth rates, averaging only 1.2 m^3/ha/year. Except for the north and northwestern areas, the western regions of the Soviet Union do not support much of the forests. Although these areas have relatively better growth rates—2.2 to 2.9 m^3/ha/year—they also have the lowest proportion of volume in mature stands (a reflection of the greater concentration of utilization in these regions) (Sutton, 1975, pp. 113–115).

Canada

Canada has two major timber-producing regions—eastern and western Canada. These two regions are separately identified, and each is formally introduced into the TSM. Eastern Canada consists of the provinces of Quebec and Ontario and the Atlantic provinces of Newfoundland, Nova Scotia, Prince Edward Island, and New Brunswick. Western Canada's production is dominated by the Province of British Columbia.

Thus far, plantation forests have played a very minor role in Canadian forestry; reforestation has been accomplished largely by natural regeneration. Total forestland in Canada amounts to about 440 million ha, of which 340 million have been inventoried; of the productive land, 260 million ha have been inventoried. Much of the noninventoried productive forestland is in eastern Canada, with Quebec accounting for 31.6 million ha and Ontario for 4.9 million ha (Bonnor, 1982, table 1, p. 4).

Although the physical volumes of Canada's forests are indisputably massive, the costs of access can be prohibitive. In a study of British Columbia's coastal forest, Williams and Gasson (1986) found that the harvestable volumes are highly dependent on price and cost considerations. At current prices accessible timber is limited and will be depleted in about a decade. However, they also found that the amount of timber that is economically harvestable would expand dramatically if real timber prices rise and/or harvesting costs were to fall.

Eastern Canada

Close to two-fifths or 130 million ha of Canada's inventoried forestland lies in eastern Canada. This area includes one-half (110 million ha) of Canada's forestland classified as "productive" (capable of producing a merchantable stand within a reasonable length of time), with 100 million ha of this being "nonreserved" (available for harvesting) and "stocked" (supporting tree growth, which includes seedlings and saplings) (Bonnor, 1982). The bulk of this forest area lies in Quebec and Ontario, which hold 48.9 million ha and 33.1 million ha, respectively.

In Canada accessibility to productive forestland is often poor. Roughly 60 percent of that land in eastern Canada can be considered a "primary supply" area where large-scale logging operations are currently in progress. Limited commercial development has taken place on about another 26 percent of the productive forestland. These areas are more remote and production is limited by the lack of transportation and processing facilities; however, they are expected to contribute to the nation's future timber supply if prices of industrial wood rise sufficiently. The remaining 14 percent of eastern Canada's forestland is suitable only for local harvests because of its occurrence in scattered patches, low yields, and other factors that limit its use to small-scale operations (Bickerstaff, Wallace, and Evert, 1981, pp. 22 and 36).

Of the eight forest regions recognized in Canada (Census and Statistics Office, 1981), four are important with respect to the forestland base of eastern Canada. The largest proportion of area is covered by the boreal forest region, which nationwide constitutes over one-half of the productive forest area. About half of this region is continuous forest, predominantly coniferous. Some of the more important softwoods of this region are black and white spruce, jack pine, and balsam fir; representative hardwoods are white birch and poplars. The lower Atlantic provinces (New Brunswick, Prince Edward Island, and Nova Scotia) are considered the Acadian forest region. Red spruce is characteristic but not exclusive; black spruce, white spruce, and balsam fir are also in abundance. Species of the Great Lakes-St. Lawrence forest region are also intermixed. This region has the highest proportion (83 percent) of productive forestland. The deciduous forest region lies in southwestern Ontario between Lakes Huron, Erie, and Ontario. However, less than 10 percent of this region is currently forested.

The forests of Canada tend to be even aged, having originated mostly after fires, insect epidemics, or harvesting. Information about the site quality of Canada's forests is incomplete, but the productive capacity and growth of Canadian forests have been estimated using the mean annual increment to maturity for natural stands of average stocking as the index. The mean annual increment at rotation age of managed stands in Canada has been found to range from about 0.3 m^3/ha for slow-growing stands on poor sites to over 10.5 m^3/ha for very good sites (Bickerstaff, Wallace, and Evert, 1981).

British Columbia

The western Canadian region modeled in the TSM is British Columbia. In terms of total land area, British Columbia is the second-largest province of Canada (only Quebec is larger) and covers about 93 million ha. Forests occupy roughly 63 million ha, and over 80 percent of this, or 52 million ha, is classified as productive forestland.

Large-scale commercial logging operations have recently been in progress in an area of about 38 million of the more productive hectares. Logging began in the coastal forest and more recently moved into interior forested areas. The relatively high rainfall and high mean temperatures found in the coastal forest make much of this area highly productive. The coastal forest has an average growth rate of about 5.0 m^3/ha, one of the highest in Canada. This region represents about 35 percent of the province's growth potential while containing only about 15 percent of the province's productive forestland.

Highly productive forestland also occurs in the Columbian region in the southeast interior of the province. This belt resembles the coastal forest except that growth is slower, the trees are smaller, and the species are more limited. In recent years this area has been increasingly logged as the industry has gradually moved inland.

Although the total forest in the province is massive, the commercial development of much of it is limited because of the limited availability of transportation and processing facilities. Nonetheless, these less exploited areas are expected to contribute to future timber supplies. However, favorable prices would be necessary to justify their exploitation, as much of the inaccessible land is in the north and at higher elevations. These forests include the subalpine forest region, which covers the uplands from the Rocky Mountains through interior British Columbia to the Pacific inlets. Spruce and lodgepole pine are predominant, and true firs are also abundant. This region is closely related to the boreal forest region, which is in the northeastern corner of the province, and to the Montane forest region of the province's interior uplands.

Industrial Wood Production in Canada

Canada's pine, spruce, and fir forests have been the backbone of its forest industry, providing the long-fibered softwoods that produce light, strong-dimension lumber and white, strong pulp and paper products. In 1979 about 95 percent of the total timber harvested in Canada was softwood. It is possible that the hardwood timber resource could become more important in the future as a source of pulp and/or paper products (USDA, Forest Service, 1982, pp. 93–94). There was a steady increase in the production of all timber products except hardwood plywood from 1950 to 1979. The greatest continued expansion has been that of softwood lumber; softwood plywood rose rapidly during the 1950s and 1960s.

British Columbia is the major producer of softwood lumber, accounting for about two-thirds or 29.5 billion m^3 of Canada's production in the late 1970s. Another one-sixth of the softwood lumber production in recent years comes from Quebec and Ontario. Softwood lumber production in the Atlantic provinces has been relatively stable, and there is limited potential for expansion (USDA, Forest Service, 1982, p. 92).

Quebec and Ontario together produce about half of Canada's wood pulp, which peaked at 22 million tons in 1974 and has remained close to that mark since then. Most of the remainder of the wood pulp comes from mills in British Columbia. The expansion of paper and board production since the late 1960s has taken place in British Columbia and the Prairie and Atlantic provinces. Throughout this time Quebec and Ontario together have maintained a relatively stable level of production at about 12 million tons per year (USDA, Forest Service, 1982, p. 93).

United States

The United States has several major forest areas, including the Pacific Coast forests, the Rocky Mountain forests, the Lake States forests, the New England forests, and the forests of the South. The major industrial wood forests are those of the Pacific Northwest region (including northern California) and the southern forest. These two regions are formally introduced into the TSM, and only these will be examined in this section.

Forest plantation establishment and intensive forest management in the United States has been actively undertaken since the mid-1950s. Early efforts (for example, the Soil Bank of the late 1950s) were justified largely on the basis of environmental considerations and were heavily subsidized. Since the mid-1960s, however, industrial plantations have been established by industrial and private landowners. In recent years over 1 million ha/year are being planted in the United States, most of them in the South and most of the remainder in the Pacific Northwest.

The Pacific Coastal Complex

This formation of forests is found in a belt that lies west of the crests of the Cascade Range and Canadian Rockies. Its northern end lies south of the Alaskan Range, and the southern portion, termed the Coast Redwood Belt, extends along the California coast to the San Francisco area. It also appears in a modified form on the western slopes of the Rockies in northern Idaho, eastern Washington, western Montana, and southeastern British Columbia. The two typical species are western hemlock and western red cedar. Other species, including redwood, Sitka spruce, and coastal silver fir, are relatively restricted in range. The most abundant and important species is Douglas fir, which is found widely.

The Pacific Northwest region of the United States constitutes one of the major timber-producing regions of the globe. Historically, timber harvested in this region has consisted largely of old growth virgin stands. However, as the old growth is drawn down, natural secondary growth forests and eventually plantation forests will become the dominant wood source.

This region consists of three quite different forests (Scott, 1980). The major producing forest is that of Washington and Oregon, west of the crest of the Cascade Mountains. This region is 80 percent forested (see table 2-8) and dominated by coastal Douglas fir. Also significant are western hemlock, western red cedar, Port-Orford cedar, and Sitka spruce as well as nonconifers such as red alder—a pioneer species that often occurs after a disturbance of forest.

Table 2-8. Pacific Northwest Forestland
(thousands of hectares)

Subregion	Forested area
Western Washington	3,963
Western Oregon	5,617
California	7,265
Idaho	5,481
Montana	5,813
Eastern Washington	3,293
Eastern Oregon	4,275

Source: U.S. Department of Agriculture (USDA), Forest Service, *Forest Statistics of the United States, 1977* (Washington, D.C., USDA, 1977).

The California region to the south is unique because of its combination of distinctive species and a Mediterranean climate. Forests extend across the full width of the state in northern California and include such species as Douglas fir, ponderosa and Jeffrey pines, and true firs as well as redwoods, cedar, and hardwoods.

The third subregion, east of the Cascades, includes the states of Idaho, western Montana, and northeastern Washington. This forest is considerably more arid than the coastal forests to the west. The dominant species include Douglas fir, true firs, lodgepole pine, and ponderosa pine. Virtually all the forested area in this subregion is mountainous; the terrain is typically rugged and steep. The climate is dominated by prevailing westerly winds along a well-developed storm track that extends the coastal climate inland to the western slopes of the Rockies. Thus, the weather is milder than would be expected at this latitude with regard to winter precipitation and summer drought.

Northern California, the forested region, is characterized by a central valley largely surrounded by mountains. The mountains surround this valley with a coast range to the west and the Sierra Nevada range to the east. To the north lie the Klamath Mountains. The eastern range is more varied; major forest zones are associated primarily with elevation.

The South

The southern forest consists of the pine region and the southern hardwood region. The southern pine region (Walker, 1980) includes the southern Atlantic and Gulf coastal plains, the Piedmont Province, and the Fall-Line Sandhills that lie between the Coastal Plain and the Piedmont. It also includes the Ozark and Ouachita mountains and the bottomlands of river courses as well as the Mississippi Delta. The 828,000-square-kilometer (km^2) Coastal Plain supports the most extensive and productive pine forests in the South. Most of the region has a humid subtropical climate characterized by high temperatures and abundant precipitation.

The total land area of the southern pine region comprises about 80 million ha in twelve states. About 60 percent of the land is forested, about 40 million ha in pine types. The growing stock of southern pines is over 2 billion m^3.

In recent years considerable tree planting has occurred in the South—almost all of it in pine. For example, in 1987 over 1 million ha of trees were planted (USDA, Forest Service, 1988a). About one-half of this planting was undertaken by the forest industry.

Major species in the Coastal Plain include loblolly pine (23 percent of the total forest), slash pine (17 percent), oak-pine (14 percent), and oak-hickory (19 percent).

The Interior Highlands refers to two different elevated provinces. The larger, lying north of the Arkansas River, is an area of broad plateaus called the Ozark plateaus. The smaller Ouachita Province lies to the south. Both have important commercial stands of short-leaf pine and upland hardwood.

The southern hardwood region includes the bottomland hardwoods along the major river courses throughout the coastal plains, the Brown Loam Bluff of mixed upland hardwood along the eastern edge of the Mississippi, and the upland hardwoods of the Appalachians. Geographically, the region extends from Pennsylvania southwest along the Ohio River to southern Illinois and eastern Oklahoma, southward to the Gulf of Mexico and Florida, and northward along the Atlantic Coast to the northern extremities of Chesapeake Bay. Of major importance are the upland hardwoods of Appalachia and the large areas of bottomland hardwoods located on the original floodplain of the Mississippi and its tributaries.

The commercial hardwood forest types in the South have been estimated to be about 55 million ha out of a total of 93 million, and the composition of the hardwood forests of the South can best be described as heterogeneous. On bottomlands these include cottonwood and willow types, cypress-tupelo types, and mixed bottomland hardwood (for example, sweet gum, water oak, and water hickory).

The Appalachian hardwood subregion is located in the unglaciated part of the eastern-central United States and encompasses most of the Appalachian Plateau, the Blue Ridge, and the Piedmont Plateau discussed elsewhere. The forest types include the mixed mesophytic forest in which dominant tree species are sugar maple, American beech, yellow poplar, northern red oak, and white oak, among others. This forest contains 40 percent of the land of the Appalachian hardwood subregion. Mixed oak is 26 percent of the commercial forest of this region. Most common species include white, northern red, and chestnut oaks; shagbark hickory; and American beech.

Oak-pine forests are located primarily in the Piedmont subregion and include 35 percent of the commercial forestland of the Appalachian hardwood subregion.

Industrial Wood Supply in the United States

The United States is the world's number-one producer of industrial wood. In the early 1980s the United States provided just under 25 percent of the world's production. Of this, 76 percent was coniferous wood, of which the United States' world share was 26 percent—again making this country the world's dominant producer. The United States is a major producer of most solidwood and fiber products. However, despite its high production, it is a net wood importer because of its very large domestic markets (Sedjo and Radcliffe, 1981).

The most recent published figures (USDA, Forest Service, 1982) indicate that in 1976, 46 percent of total U.S. timber harvests were in the South and another 31 percent were in the Pacific Northwest. However, production has been shifting to the South, reflecting the drawdown of old growth in the West and the favorable growing and other conditions in the South. Although concern has been expressed over the adequacy of future timber supply in the South (Brooks, 1985), as noted, substantial amounts of plantation establishment and tree planting are currently under way in the South.

Latin America

An Overview

Latin America's natural forests cover large areas from northern Mexico to near the Antarctic in Argentina and Chile. The rain forest of South America centered in the Amazon Basin is the greatest continuous rain forest in the world, accounting for half of all moist tropical forests.

The altitude range of Latin American forests is from sea level to 3,500 meters. As a result of this wide latitude and altitude range, the forests are of many very different types. The natural pine forests, growing mainly in Central America and Mexico, and the mixed tropical and temperate hardwood forests spread over the entire region should be distinguished. The natural araucaria forests, which grow mainly in southern Brazil, have been heavily exploited and are probably of little economic importance in the future.

The total natural forests cover about 720 million ha, or 36 percent of the region's total land area (table 2-9). Although about three-quarters of this is considered to be of a more closed forest type that could be industrially utilized, no more than 3 percent of all forest is coniferous, and the remainder is mainly mixed tropical hardwood forest. Brazil, with its huge tropical forests, has the region's main natural hardwood resource (McGaughey and Gregersen, 1983).

Despite its massive areas of forestland and huge volumes of standing timber, Latin America is a relatively minor source of industrial wood. Hence Latin America is not formally introduced into the TSM as a separate region. However, many of the countries represented in the model's emerging region

Table 2-9. Latin American Forest Areas, 1980
(land area in thousands of hectares)

Subregion	Total land area	Total forest area	Percentage of land area	Area of natural productive forest		Annual deforestation	Total forest plantation	Area of industrial plantations		Annual planting area
				Coniferous	Broadleaf			Coniferous	Broadleaf	
Mexico	197,255	46,250	23.5	11,720	12,580	530	159.0	37.0	35.0	7.8
Central America	50,862	18,679	36.7	2,512	11,682	382	25.4	15.8	9.6	3.9
Caribbean	56,435	44,511	78.9	277	34,960	21	48.8	26.1	16.0	8.1
Brazil	851,196	357,480	42.0	280	300,910	1,360	3,855.0	1,232.0	741.0	158.0
Andean	446,311	206,210	46.2	185	142,975	1,535	372.4	181.8	115.6	26.8
Southern Cone	412,727	46,605	11.3	820	28,040	155	1,453.0	874.8	410.1	93.2
Total Latin America	2,014,786	719,735	35.7	15,794	531,147	3,983	5,913.6	2,367.5	1,327.3	297.8

Source: Stephen E. McGaughey and Hans M. Gregersen, eds., *Forest-Based Development in Latin America* (Washington, D.C., Inter-American Development Bank, 1983). Reprinted with permission.

are found in Latin America, Chile and Brazil being the more important. In terms of land area currently or projected to be in plantation forests in the emerging region, Latin America dominates.

Most of the region's natural hardwood forest area (about 80 percent) is state owned, whereas plantations are mainly private; the distribution of the natural pine forest is about 60 percent public and 40 percent private. There is a trend toward increasing private ownership, as more and more state-owned land is being claimed by settlers.

The total growing stock in the productive natural forest is estimated as 1,180 million m^3 (about 75 m^3/ha) for the coniferous forest, and 79,110 million m^3 (about 150 m^3/ha) for the hardwood forest. The total roundwood removals in Latin America from natural forests are about 350 million m^3/year. This is only 0.4 percent of the growing stock, which indicates that Latin America's forests are underutilized. However, many areas are badly overexploited, as cutting has concentrated on the most accessible areas and on a limited number of commercial species. The natural forests are either not being utilized at all (for industrial wood production) or are often exploited by a single phase of logging followed by burning and cattle grazing.

More than four-fifths of the total wood production is for local fuelwood consumption; such production thus makes an important contribution to the basic living needs of the rural population. Only about 53 million m^3/year are utilized industrially, most of which (46 million m^3) is sawlogs and veneer logs.

Pulpwood is extracted mainly from plantations (16 million m^3/year). Natural coniferous forests produce no more than 5.2 million m^3 of pulpwood per year. Because the logging costs in tropical hardwood forests have generally been very high, and because of technical difficulties associated with pulping mixed species, only smaller volumes (1.2 million m^3/year) of pulpwood have been extracted from them. Because of the expected increase in transport costs and relatively low production costs of plantation wood, it is unlikely that the share of pulpwood derived from natural hardwood forests will significantly increase in the region. Instead the main future industrial use of tropical hardwood will be for sawnwood and panels. In natural pine forests (in Central America and Mexico), the volume of the wood utilized for pulping is likely to increase with growing demand. Industrial utilization may be limited in some regions by the growing local demand for fuelwood.

Plantations

Latin America figures heavily in this study's assessment and projections of future production from forest plantations in the emerging region. About 6 million ha had been planted there by 1980; however, somewhat less than that amount was considered to be in plantations of industrial size and suitable location.

Many of the region's plantations are less than 10 years old. About two-thirds of the total industrial plantations are coniferous species, mainly tropical pines, and one-third is fast-growing hardwood species, mainly eucalyptus. Argentina, Brazil, and Chile account for 85 percent of the region's total industrial plantation area.

It is estimated that Latin America's industrial plantation area will be tripled by the year 2000, covering about 11 million ha. The biological potential is estimated to be almost 30 million ha (McGaughey and Gregersen, 1983, p. 80). Annual planting rates are projected to increase no more than 10 percent in softwoods but by almost 80 percent in hardwoods between 1980 and 2000.

The foregoing projections exclude fuelwood plantations. However, in Brazil, for example, an additional annual planting area of 100,000 to 200,000 ha is intended for the production of fuelwood and charcoal on an industrial scale. It is not expected that fuelwood plantations would seriously limit the availability of land for industrial plantations.

Plantation yield varies considerably by species in view of the wide range in climate, elevation, and soil. In Chile, Monterey pine can reach as much as 20 to 30 m^3/ha/year, but in Argentina and southern Brazil the annual growth of loblolly pine varies from 15 to 25 m^3/ha/year. In Brazil there are many fast-growing eucalyptus plantations where 30 to 35 m^3/ha or even more can be expected. However, average figures should be much lower, as many plantations are far from optimal with respect to site and species selection and probably also stocking.

It is estimated that by the year 2000 the wood production of Latin American industrial plantations will be almost four times higher than it is now and will account for about 50 percent of the total industrial wood supply. This switch from natural forests to plantations will take place mainly in Brazil and the Southern Cone. In Mexico and Central America, natural pine forests will continue to be important resources at the end of the century if they can be brought under more intensive management and control than they are currently.

The plantation wood is mainly pulpwood (70 percent), but an increasing output of sawlogs and veneer logs is anticipated because of the increasing demand for these valuable products, which are more and more difficult to obtain from natural hardwood forests. In Chile, by contrast, most of the plantation wood is sawlogs, although this could change if more capital becomes available for a large expansion of pulp and paper production.

Asia

An Overview

Asian forests stretch from Hokkaido in northern Japan through the moist tropical forests of Southeast Asia to the dry, deciduous forest of the western

Indo-Pakistan region of Southwest Asia. These forests include China's forests, which range from tropical rain forests in the south through temperate broadleaves to the boreal coniferous forests of the far north. The majority of these forests are tropical nonconifer forests.

The vast areas of Asia's tropical forest account for about one-third of the world's moist tropical forests and include the tropical rain forests of Malaysia, Indonesia, the Philippines, Papua New Guinea, and the Solomon Islands, as well as the monsoonal forests of continental Southeast Asia, which include forestlands in Burma, Indochina, Thailand, and parts of India.

Although within Asia several countries are significant producers of industrial wood, the only Asian forests formally introduced into the TSM are the forests of the Asia-Pacific region. These forests, dominated by the dipterocarp species, run from Malaysia to Papua New Guinea; the largest forested areas are in Indonesia.

Japan is the world's leading importer of industrial wood but also a major producer in its own right, supplying about 40 percent of its domestic raw wood requirements in recent years. However, although Japan is a major producer of industrial wood, the massive domestic requirements make it almost certain that Japan will continue to be a major net wood importer into the indefinite future.

About 90 percent of the production of industrial wood in the Indian subcontinent and Sri Lanka is hardwood. Almost all the production is consumed domestically.

Certain Asian countries have been successful with major reforestation and afforestation programs. Japan has established over 10 million ha in forest plantations as the result of a major effort to improve its depleted forests after World War II. More recently, major afforestation programs have been undertaken in China and South Korea. As noted above, China has afforested 20 to 40 million ha. Korea's program, while more modest, has succeeded in the afforestation of about 4 million ha.

More modest efforts at afforestation are under way in Nepal, Pakistan, Bangladesh, and India. These efforts often combine environmental protection goals with those of fuelwood and sometimes industrial wood production.

Other modest plantation activities are under way in Malaysia, Indonesia, and the Philippines. Also, specialized wood (for example, teak) has been grown in plantation forests in some parts of Asia for many decades.

Forest Resources of Tropical Asia

The sixteen countries of tropical Asia stretch from India to Papua New Guinea and may also be extended logically to the Solomon Islands. In 1980 the total area and forests in the sixteen countries of tropical Asia was 445 million ha, or about 47 percent of the land's surface (FAO, 1981). The region can be divided into four subregions: South Asia, continental Southeast

Asia, insular Southeast Asia, and Papua New Guinea. Table 2-10 presents estimates of the forest area by subregion and country.

The forests of the Asia-Pacific region are clearly the most important industrial forests within the tropical regions, accounting for over 50 percent of the total industrial wood harvested from tropical forests. The tropical hardwood resources of the tropical Asia-Pacific region can be separated into four forest regions: the dipterocarp areas of the mainland and insular Southeast Asia, the nondipterocarp area of the southwest Pacific Islands, the teak forest area of Burma and Thailand, and the Indian subcontinent plus Ceylon.

Commercially, the dipterocarp area is the most important. It includes the Philippines, Malaysia, Indochina, and Indonesia west of the "Wallace Line" that separates the Malukus, Lesser Sunda Islands, and Irian Jaya from the rest of Indonesia (Takeuchi, 1974). Timber trade in the tropical Asia-Pacific region has been dominated by the export of logs of the *Dipterocarpus* and

Table 2-10. Areas of Closed Forest in Tropical Asia by Subregion and Country, 1980

Subregion/country	Hectares (thousands)	Percentage of region
South Asia	60,653	19.86
Bangladesh	927	0.30
Bhutan	2,100	0.69
India	51,841	16.97
Nepal	1,941	0.64
Pakistan	2,185	0.72
Sri Lanka	1,659	0.54
Continental Southeast Asia	65,904	21.57
Burma	31,941	10.46
Kampuchea	7,548	2.47
Laos	8,410	2.75
Thailand	9,235	3.02
Viet Nam	8,770	2.87
Insular Southeast Asia	144,723	47.37
Brunei	323	0.11
Indonesia	113,895	37.28
Malaysia	20,995	6.87
(Peninsular)	(7,578)	(2.48)
(Sabah)	(4,997)	(1.63)
(Sarawak)	(8,420)	(2.76)
Philippines	9,510	3.11
Papua New Guinea	34,230	11.20
Total	305,510	32.33

Source: Food and Agriculture Organization of the United Nations (FAO), *Forest Resources of Tropical Asia* (Rome, FAO, 1981), p. 40. Reprinted with permission.

Shorea genera, mainly from the Philippines, Malaysia, and Indonesia. The resources of Indochina are plentiful and marketable; however, the lack of political stability has prevented their development.

The forest resources in the southwest Pacific Islands, though rich in volume, are of nondipterocarp species, most of which are not yet known in major markets. Currently both the Solomon Islands and Papua New Guinea are modest exporters of logs.

Burma and Thailand, together with the island of Java, have traditionally been the major sources of teakwood, with Burma as the dominant supplier.

Africa

Africa has some 800 million ha or 27 percent of its land area in forest, but only 190 million ha are closed forest. Thus, most of Africa's forests are open woodland, such as savannah, or in brush and scrub (table 2-2). Much of the closed forest is tropical rain forest, which occurs on the south coast of West Africa, in central Africa focused on the Zaire River Basin, and as a fringe forest on the east coast of Madagascar.

The West African rain forest has been heavily logged for export markets and substantial deforestation has occurred. By contrast, the central African forest is largely undisturbed. Although the central African forests present the best opportunity for commercial exploitation, this area is also becoming a point of concern for environmentalists anxious about preserving these tropical forests.

Plantation forests are beginning to appear in a modest way in parts of Africa; eucalyptus and gmelina species are used in parts of tropical Africa. Perhaps the most successful industrial plantations, however, are the pine plantations of South Africa.

Except for South Africa, which is incorporated in the TSM as part of the emerging region, Africa is not formally represented in the model. It appears only in aggregation as part of the "nonresponsive" region.

Oceania

The native forests of Australia and New Zealand, having developed apart from the forests of Asia, show remarkable differences from those of the southernmost part of the Indo-Malaysian region of western Indonesia and Southeast Asia. Many of the tree species indigenous to the region are unique to it or are found elsewhere only south of the equator in South Africa and South America (Johnson, 1981). The largest group is the eucalyptus, a highly adaptable genus of more than 500 species. This genus has demonstrated remarkable growth in some environments and is being widely used in Europe and South America as an exotic in intensively managed plantation forestry.

With the exception of the indigenous eucalyptus forests of Australia, the industrial wood potential of Oceania appears to reside primarily in its exotic plantation forests, which are largely pine. The plantation sector of these countries is introduced formally into the TSM as a part of the emerging region.

Australia

Vast areas of Australia are desert or semi-desert, and closed forests flourish only in the more humid uplands and coastal areas of southwestern and eastern Australia. Two distinct forest formations are recognized: the complex tropical rain forest of Queensland in the northeast of the continent and the sclerophyll forests in areas of low rainfall. The eucalypts dominate 95 percent of Australia's forests where they have become adapted to a wide range of conditions. Although some valuable softwoods are indigenous to the temperate regions of Australia, the indigenous conifers have been supplemented with extensive plantation of exotic species to provide softwood lumber and long fiber for pulping.

Australia continues to be a major wood importer but has also become a significant source of wood chips for Japan. This nation appears to have great potential to expand its industrial forests and their production.

New Zealand

Although New Zealand is isolated in the Pacific, its trees are similar to those of Australia. However, New Zealand's varying climate and terrain encourage a patchwork distribution. About 26 percent of New Zealand (7 million ha) is forested. Indigenous species compose most of the resource, but their contribution to the harvest of industrial wood has steadily been diminishing since 1950. Current timber production is based largely on exotic softwood species, which account for about 1.1 million ha.

Prior to 1900 the domestic industry relied entirely on indigenous species. During the late nineteenth century and early part of the twentieth century, the government began an experimental tree-planting program, and by 1939 over 300,000 ha of plantation had been established. The maturity of these forests after 1950 resulted in the rapid development of a wood-processing industry, and a subsequent search for markets resulted in New Zealand exports to foreign markets, particularly Australia and Japan.

New Zealand continues to expand its forest plantation lands at the rate of about 50,000 ha/year. The age distribution of the current inventory suggests very large increases in mature trees in the 1990s and especially after 2000, with the probable implication that exports of wood and wood products will increase dramatically.

SUMMARY

This chapter has presented a broad overview of the world's forests, focusing chiefly on the industrial wood-producing regions, particularly the seven "responsive" regions that are formally part of the TSM. We have discussed the process of transition from natural forests to plantation forests, the existing forest volume and its suitability for utilization as industrial wood, and the timber-growing characteristics of each region. The entire discussion is set in the context of an environment in which the world's forests are viewed as not static but rather as dynamic in time.

Historically, the forests have changed as pressures have been placed on them by expanding human populations. The need for land for nonforest purposes and demands on the timber resource have combined to reduce the forest resource. A countervailing force in favor of the forest, however, is found in the recognition by humans of the importance of the forest for environmental protection. Humans have acquired the ability to plant and manage forests for their timber value. In addition, in many places, especially in the temperate regions, marginal agriculture has ceased and the forest has been resurrected, often through natural processes. Although a tension often exists between forest uses and other land uses, in much of the temperate world a balance has been struck that has resulted in the stabilization of the land area in forest.

In the tropical regions of the world, in contrast, the forest area continues to contract. Although this phenomenon has many undesirable features, from the point of view of industrial wood and wood fiber production the impacts are likely to be modest at best. The regions of the globe from which most of the world's industrial wood is harvested are largely temperate-climate areas where in general the land area in industrial forest production has stabilized.

REFERENCES

Adams, Davis L. 1980. "The Northern Rocky Mountain Region," in John W. Barrett, ed., *Regional Silviculture of the United States* (New York, Wiley).

Barr, Brenton M. 1979. "Soviet Timber: Regional Supply and Demand, 1970–1990," *Arctic* vol. 32, no. 4 (December) pp. 308–328.

Bickerstaff, A., W. L. Wallace, and F. Evert. 1981. *Growth of Forests in Canada, Part 2: A Quantitative Description of the Land Base and the Mean Annual Increment.* Information Report P1-X-1 (Ottawa, Ontario, Canadian Forestry Service, Environment Canada).

Bonnor, G. M. 1982. *Canada's Forest Inventory, 1981* (Ottawa, Ontario, Forestry

Statistics and Systems Branch, Canadian Forestry Service, Department of the Environment).

Brooks, David. 1985. "Public Policy and Long-Term Timber Supply in the South," *Forest Science* vol. 31, no. 2 (June) pp. 342–357.

Census and Statistics Office. 1981. *Canada Year Book, 1980–1981* (Ottawa, Ontario, Supply and Services—Canada).

Clark, Thomas D. 1984. *The Greening of the South* (Lexington, Ky., University Press of Kentucky).

Economic Commission for Europe/Food and Agriculture Organization of the United Nations (ECE/FAO). 1976. "European Timber Trends and Prospects 1950 to 2000," *Timber Bulletin for Europe,* supp. 3 to vol. XXIX (New York, United Nations).

———. 1985. *The Forest Resources of the ECE Region* (Geneva, ECE/FAO).

Food and Agriculture Organization of the United Nations (FAO). 1978. "Development and Investment in the Forestry Sector," FO:COFC-78/2 (Rome, FAO).

———. 1980. *Yearbook of Forest Products 1969–1980* (Rome, FAO).

———. 1981. *Forest Resources of Tropical Asia* (Rome, FAO).

Food and Agriculture Organization of the United Nations/United Nations Environmental Program (FAO/UNEP). 1982. "Tropical Forest Resources," FAO Forestry Paper no. 30 (Rome, FAO/UNEP).

Gill, Gerard J. 1985. "The Control and Degradation of Land Resources Under Intense Population Pressure: The Case of Bangladesh," paper presented at conference on "Managing Renewable Resources in Asia," June 24–28, Sapporo, Japan.

Helms, John A. 1980. "The California Region," in John W. Barrett, ed., *Regional Silviculture of the United States* (New York, Wiley).

Holowacz, J. 1974. "Soviet Forest Resource Analyzed," *World Wood* vol. 15, no. 13 (December) pp. 13–15.

———. 1985. "Forests of the USSR," *Forestry Chronicle* vol. 61, no. 5 (October) pp. 366–373.

Inter-American Development Bank (IDB). 1982. "Forest Industries Development Strategy and Investment Requirements in Latin America: Technical Report No. 1," paper prepared for IDB Conference on Financing Forest-Based Development in America, June 22–25 (Washington, D.C., IDB).

Johnson, Hugh. 1981. *The International Book of the Forest* (New York, Simon and Schuster).

Lanly, J. P., and J. Clement. 1979. "Present and Future National Forest and Plantation Areas in the Tropics," *Unasylva* vol. 31, p. 123.

Manthy, Robert S. 1978. *Natural Resource Commodities—A Century of Statistics* (Baltimore, Md., The Johns Hopkins University Press for Resources for the Future).

McClintock, Wayne, and Nick Taylor. 1983. "Pines, Pulp and People," Information Paper no. 2 (Christ Church, New Zealand, University of Canterbury and Lincoln College, Centre for Resource Management).

McGaughey, Stephen, and Hans M. Gregersen, eds. 1983. *Forest-Based Development in Latin America* (Washington, D.C., Inter-American Development Bank).

Menzies, Nick. 1985. "Land Tenure and Resources Utilization in China: An Historical Perspective," paper presented at conference on "Managing Renewable Resources in Asia," June 24–28, Sapporo, Japan.

Persson, Reider. 1974. *World Forest Resources: Review of the World's Forest Resources in the Early 1970s*, no. 17 (Stockholm, Royal College of Forestry).

Plochman, Richard. 1984. "Air Pollution and the Dying Forests of Europe," *American Forests* vol. 90, no. 6 (June) p. 17.

Reed, F. L. C., and Associates. 1978. *Forest Management in Canada,* vol. 1. Information report FMR-X-102 (Vancouver, British Columbia, Forest Management Institute).

Scott, David R. M. 1980. "The Pacific Northwest Region," in John W. Barrett, ed., *Regional Silviculture of the United States* (New York, Wiley).

Sedjo, Roger A. 1980. "Forest Plantations in Brazil and Their Possible Effect on World Pulp Markets," *Journal of Forestry* vol. 78, no. 11 (November) pp. 702–705.

———. 1983. *The Comparative Economics of Plantation Forestry: A Global Assessment* (Washington, D.C., Resources for the Future).

Sedjo, Roger A., and Marion Clawson. 1983. "Tropical Deforestation: How Serious?" *Journal of Forestry* vol. 81, no. 12 (December) pp. 792–794.

———. 1984. "Global Forests," in Julian Simon and Herman Kahn, eds., *The Resourceful Earth* (New York, Basil Blackwell).

Sedjo, Roger A., and Samuel J. Radcliffe. 1981. *Postwar Trends in U.S. Forest Products Trade* (Washington, D.C., Resources for the Future).

Sutton, W. R. J. 1975. "Forest Resources of the USSR: Their Exploitation and Potential," *Commonwealth Forestry Review* vol. 54, no. 160 (June) pp. 110–138.

Takeuchi, Kenji. 1974. "*Tropical Hardwood Trade in the Asia-Pacific Region,*" World Bank Staff Occasional Paper no. 17 (Washington, D.C., World Bank).

Thirgood, J. V. 1981. *Man and the Mediterranean Forest—A History of Resource Depletion* (London and New York, Academic Press).

United Nations Industrial Development Organization (UNIDO). 1983. "The USSR Forest and Woodworking Industries," Sectoral Working Paper Series no. 7 (Vienna, UNIDO).

U.S. Department of Agriculture (USDA), Forest Service. 1977. *Forest Statistics of the United States, 1977* (Washington, D.C., USDA).

U.S. Department of Agriculture (USDA), Forest Service. 1982. "An Analysis of the Timber Situation in the U.S.—1952-2030," *Forest Resource Report no. 23* (Washington, D.C., USDA).

U.S. Department of Agriculture (USDA), Forest Service. 1988a. "The South's Fourth Forest: Alternatives for the Future," Forest Resource Report no. 24 (June) (Washington, D.C., USDA).

U.S. Department of Agriculture (USDA), Forest Service. 1988b. "An Analysis of the Timber Situation in the United States: 1989–2040," Part 1, "A Technical Document Supporting the 1989 RPA Assessment" (draft) (Washington, D.C., USDA).

Vassiliev, P. V. 1961. "USSR Forest Resources and Features of Their Inventory," *Unasylva* vol. 15, no. 3, pp. 119–124.

Walker, Laurence C. 1980. "The Southern Pine Region," in John W. Barrett, ed., *Regional Silviculture of the United States* (New York, Wiley).

Williams, Douglas, and Robert Gasson. 1986. "The Economic Stock of Timber in the Coastal Region of British Columbia" (Vancouver, University of British Columbia, Forest Economics and Policy Analysis Program).

World Bank. 1982. "The People's Republic of China: Environmental Aspects of Economic Development," Office of Environmental Affairs, Project Advisory Staff (Washington, D.C., IBRD).

World Resources Institute. 1986. *World Resources 1986* (New York, Basic Books).

World Wood. 1981. *1981 World Wood Review* vol. 22, no. 8.

3

Long-Term Price and Consumption Behavior

This chapter provides a context for viewing the problem of long-term timber supply and introduces some of the problems addressed and concepts utilized later in this study. In addition, the historical trends identified here provide the basis for the projections of future demand growth used in the formal Timber Supply Model. First, the historical experience of wood and wood resource real prices is examined as an indicator of resource scarcity and intertemporal changes in relative scarcity. This examination is undertaken in the spirit of earlier studies by Barnett and Morse, Potter and Christy, and Manthy, and several generalizations and implications are drawn from the examination. Next, four hypotheses are offered to explain the historical behavior of prices and harvests and particularly changes observed in the pattern. These hypotheses are not mutually exclusive. Finally, the various pieces are combined to provide an integrated view that explains price and production behavior within the context of concepts common to resource economics. This integrated view is generally consistent with the more formal TSM developed in chapter 4. It presents the reader with a preliminary glimpse of the problems addressed by the formal model and many of the concepts embodied in the model.

Although the investigation is global, limitations in worldwide data and the greater availability of U.S. data (particularly price data) cause much of the analysis to focus on the United States. However, the trends for the United States can be generalized to the world. This is so because in an integrated world market where trade barriers are modest and fairly stable and where significant interregional trade flows exist, the "law of one price" dictates that the direction of changes in prices in the several regions be in the same direction. Available cross-country empirical data tend to support this approach. Although it may be argued that world trade in forest products was not well integrated before World War II, this is clearly not true for the postwar period. Furthermore, trade barriers have been relatively modest for forest products in the postwar period (Bourke, 1986, p. 1). This last point is important, as the focus of this section and the demand growth assumed in the TSM are developed largely on the basis of the postwar experience.

HISTORICAL PRICE AND CONSUMPTION BEHAVIOR

Global Experience

Tables 3-1 and 3-2, which give data on worldwide industrial wood consumption and production for the post-1950 period, provide much of the rationale for the assumptions used later as to the anticipated rate of growth of future

Table 3-1. Worldwide Annual Growth in Consumption of Industrial Wood, 1950–1985

Period	Production/consumption (percentage)
1950–1960	3.54
1960–1970	2.20
1970–1980	1.10
1970–1985	0.94
1950–1980	2.30
1950–1985	2.01

Source: Data from selected issues of Food and Agriculture Organization of the United Nations (FAO), *Yearbook of Forest Products* (Rome, FAO).

Table 3-2. Worldwide Annual Growth of Industrial Wood Consumed in Pulp Products, 1964–1985

Period	Production/consumption (percentage)
1964–1974	4.16
1970–1980	1.60
1970–1985	1.40
1964–1980	2.98
1964–1985	2.53

Note: This series does not begin in 1950, as the reported data do not allow a separate breakout of industrial wood for pulp products before 1964.

Source: Data from selected issues (on pulpwood and particles) of Food and Agriculture Organization of the United Nations (FAO), *Yearbook of Forest Products* (Rome, FAO).

industrial wood demand. For the consumption of all industrial wood, the thirty-five-year average annual growth rate was 2.01 percent. However, the rate of growth was far from stable over the period; rather, it declined persistently over the postwar decades. The annual growth rate from 1970 through 1985 was only 0.94 percent. Table 3-2 presents a similar situation for the more rapidly growing pulpwood for a twenty-one-year period beginning in 1964. Again the rate of growth worldwide declined systematically over time. Although the twenty-one-year growth period of pulpwood averaged 2.53 percent, the annual growth rate of the 1970–1985 period was only 1.4 percent. In the case of both total industrial wood and pulpwood, the worldwide rates of consumption growth clearly declined over the post-1950 period.

Tables 3-3 and 3-4 provide information on the behavior of the real price of industrial wood in the United States during the post-1950 period. The data for the United States indicate that while real prices exhibited a high degree of volatility over the post-1950 period, especially during the decade of the 1970s, overall the real price of industrial wood had increased only very modestly in the mid-1980s from its level in 1950.

As we have noted, the "law of one price" suggests that industrial wood real price movements in the United States should mirror movements elsewhere, and in fact U.S. real price behavior *is* consistent with the behavior of the real prices of industrial wood elsewhere in the world. For example, the

Table 3-3. Annual Growth of Consumption and Real Prices of Industrial Wood in the United States, 1900–1985
(percentage growth)

Consumption/price growth	1900–1950	1950–1985	1900–1985
All industrial wood consumption	0.60	1.11	0.81
Real price growth[a]	1.37	0.34	1.02
Pulpwood consumption	4.74	2.17	3.67
Pulpwood real price growth	0.13	0.18	0.15
Sawlogs real price growth	2.86	0.50	1.88

Source: Indexes constructed from Robert S. Manthy, *Natural Resource Commodities—A Century of Statistics* (Baltimore, Md., The Johns Hopkins University Press for Resources for the Future, 1978); Marion Clawson, "Forests in the Long Sweep of American History," *Science* vol. 204, no. 4398 (June 1979) pp. 1168–1174; and "U.S. Timber Production, Trade, Consumption, and Price Statistics 1950–1986," misc. pub. no. 1453 (Washington, D.C., U.S. Department of Agriculture, Forest Service, 1988), p. 13. Real prices for the period from 1900 to 1950 are from Manthy (1978). Post-1950 prices are for sawlogs, as provided in Clawson (1979), averaging South and Douglas fir regions.

[a]The real price growth for "all industrial wood" is the unweighted average of the sawlog and pulpwood real prices. The general trends for logs are confirmed by the stumpage real price trends found by Russell B. Milliken and Frederick W. Cubbage, "Trends in Southern Pine Timber Price Appreciation and Timberland Investment Returns, 1955 to 1983," Research Report 475 (Athens, University of Georgia, College of Agriculture Experiment Station, 1985).

Table 3-4. Annual Growth of Consumption and Real Prices of Industrial Wood and Pulpwood in the United States, 1950–1985
(percentage growth)

	All industrial wood		Pulpwood		Sawlogs
Period	Consumption	Price	Consumption	Price	Price
1950–1960	0.23	0.48	3.22	1.2	0.24
1960–1970	1.68	−0.35	2.91	−1.5	−1.54
1970–1980	0.45	4.52	1.05	2.1	6.93
1970–1985	1.37	0.05	0.93	0.6	−0.50
1950–1985	1.11	0.34	2.17	0.2	0.50

Source: Indexes constructed from Robert S. Manthy, *Natural Resource Commodities—A Century of Statistics* (Baltimore, Md., The Johns Hopkins University Press for Resources for the Future, 1978); Marion Clawson, "Forests in the Long Sweep of American History," *Science* vol. 204, no. 4398 (June 1979) pp. 1168–1174; and "U.S. Timber Production, Trade, Consumption, and Price Statistics, 1950–1986," misc. pub. no. 1453 (Washington, D.C., U.S. Department of Agriculture, Forest Service, 1988), p. 13. Real prices for the period from 1900 to 1950 are from Manthy (1978). Post-1950 prices are for sawlogs, as provided in Clawson (1979), averaging South and Douglas fir regions.

The real price growth for "all industrial wood" is the unweighted average of the sawlog and pulpwood real prices. The general trends for logs are confirmed by the stumpage real price trends found by Russell B. Milliken and Frederick W. Cubbage, "Trends in Southern Pine Timber Price Appreciation and Timberland Investment Returns, 1955 to 1983," Research Report 475 (Athens, University of Georgia, College of Agriculture Experiment Station, 1985).

findings of the study *European Timber Trends and Prospects to the Year 2000 and Beyond* (ECE/FAO, 1986), which examines the period from 1964 into the mid-1980s, reveal the same absence of an upward or declining real price trend. While finding strong fluctuations in prices, most notably in the 1970s, the ECE/FAO study concludes that "no strong underlying long-term [price] trends were identified of such a nature that they could be expected to continue into the future" (p. 160). Furthermore, an examination of another publication, "Forest Products Prices: 1963–1982" (FAO, 1983), reveals the same absence of a strong long-term trend in real prices for industrial wood in a variety of other regions.

The finding of no significant long-term real price changes in recent decades for industrial wood in many of the world's major markets implies that the increase in consumption experienced over that period is approximately equal to the outward shift in the worldwide demand function for industrial wood over that period. This finding provides a useful basis for estimating future shifts in the worldwide demand function for industrial wood.

The United States' Experience

Table 3-3 subdivides the first eighty-five years of the twentieth century in the United States into two subperiods from 1900 to 1950 and from 1950 to

1985. The differences between these two periods are striking. During the 1900–1950 period, industrial wood consumption experienced an average annual growth rate of 0.60 percent and an annual real price rise averaging 1.37 percent. In contrast, the more recent period from 1950 to 1985 showed a significantly more rapid rate of growth in consumption (1.11 percent annually) but an almost negligible growth rate in real price (0.34 percent annually). For pulpwood (a subset of total industrial wood), the rate of growth in consumption was much more rapid, averaging 4.74 percent over the fifty-year period and declining to 2.17 percent in the post-1950 period. Also, the real price of pulpwood experienced almost negligible increases (0.15 percent annually) over the entire eighty-five-year period. For sawlogs, although the growth rate of sawlog real prices was a significant 2.86 percent per year for the 1900–1950 period, the post-1950 real price growth rate fell to a very modest 0.50 percent.

At least three surprises appear in these findings for the United States. First, for "all industrial wood," the more modest period of consumption growth of the pre-1950s exhibited substantially more rapid increases in real prices than did the more rapid consumption growth period of the post-1950s. Second, although the United States is viewed as a relatively rapidly growing market, the growth in U.S. consumption lagged behind worldwide growth for all comparable post-1950 periods except the most recent 1970–1985 period. Finally, total U.S. consumption did not exhibit the neat, systematic decline in growth that was characteristic of worldwide consumption for the post-1950 period.

Table 3-4 examines the postwar period of 1950 to 1985 in greater detail, focusing on the changes in price and consumption by decade in the United States. For "all industrial wood" considerable variation occurs across decades; the 1970s show a strong increase, while the rest of the period is characterized by small price changes and some actual declines. In fact, if the twelve-year period from 1967 to 1979 is isolated, a much more dramatic increase in prices and consumption can be observed, with both pulpwood and sawlogs showing rapid real price rises. However, by 1985 the sharp real price rises of the 1970s have been largely offset by subsequent real declines, and the real price increase for the entire thirty-five-year period is dramatically reduced to an annual rate of only 0.5 percent. For sawlogs particularly, the post-1970 declines in real prices largely offset the earlier increases, and for the thirty-five-year period, sawlog real prices experienced a very modest overall increase.

These findings of the absence of a strong trend for industrial wood real prices for the 1950–1985 period tend to confirm the earlier observations of Manthy (1978, p. 14) and are substantiated by numerous other data. For example, Forest Service data (USDA, Forest Service, 1988) indicate that real lumber prices were roughly the same in the mid-1980s as they were in the early 1950s. Similar real price behavior is found for pine stumpage in the

South (Milliken and Cubbage, 1985) and Douglas fir stumpage in the Pacific Northwest (USDA, Forest Service, 1988). These findings call into question the common assumption among foresters and others that industrial wood supplies are consistently being outrun by increases in demand and that the market manifestation of this trend is rising real prices. The reality is that the real prices of most types of industrial wood have exhibited only a modest (if any) long-term upward trend over the postwar period from 1950 to 1985 and that any price growth observed was certainly much less than had occurred previously.

Although total industrial wood consumption in the United States showed considerable variation across decades for the postwar period, the average annual growth for this period was significantly above that of the 1900–1950 period. In contrast, pulpwood consumption (including chips and residuals) showed a lower growth rate in the postwar period than in the 1900–1950 period. Pulpwood consumption exhibited a consistent declining trend in its growth over the 1950–1985 period, as its annual rate of growth fell from 3.22 percent in the 1950s to 1.05 percent in the 1970s and to 0.93 percent for the 1970–1985 period. Thus, over the thirty-five-year period the growth rate of industrial wood total consumption and the growth rate of pulpwood moved toward convergence. This is not really surprising when one recognizes that pulpwood's share of total consumption is increasing markedly over time; pulpwood constituted a negligible fraction of industrial wood in 1900 and only about 23 percent of all industrial wood in 1950, but it rose to 34 percent of industrial wood by 1980.

SOME GENERALIZATIONS AND IMPLICATIONS

What generalizations and implications can be drawn from the data presented? First, the worldwide rate of growth of industrial wood consumption has exhibited a substantial decline during the post-1950 period. Second, the U.S. real price, and by inference the world price, of most industrial wood is only modestly higher at the end of the thirty-five-year post-1950 period than it was at the beginning. This contrasts dramatically with the long-term real price rises experienced for industrial wood—sawlogs and lumber—in the 1900–1950 period. Third, the shift in the worldwide structure of industrial wood from solidwood to pulp and fiber products, which began about 1900, has continued in the post-1950 period, both worldwide and in the United States. Fourth, in recent years the consumption growth rates have tended to converge across products and regions. Thus, the growth rates are becoming more similar for pulpwood and all industrial wood, and the U.S. growth rate is becoming more similar to that worldwide.

What is the significance of these facts? One implication of the stabilization of the long-run real price for industrial wood is that for the postwar period as a whole, supply and demand shifts have tended to be largely offsetting. In the post-1950 period this has been the case not only for pulpwood but also for solidwood such as sawlogs. This situation contrasts with the 1900–1950 period when demand tended to outrun supply, thus resulting in rising real prices for most wood resources. Put another way, the lack of a strong upward price trend suggests that the half-century (and longer) phenomenon of increasing economic scarcity of industrial wood appears largely to have ceased after 1950, although shorter-term fluctuations surely do remain.

A second important point is that the rapid increases in consumption that characterized the early postwar period have tended to moderate substantially over time. This has implications for any set of demand projections that may be built into a model assessing the long-term timber supply. Finally, the structural shift of the industry away from solidwood products and to pulp products is significant in that pulpwood is having a greater effect on the entire industrial wood sector. Should this shift continue, demand for pulp products will increasingly become the sector's driving force. This in turn has implications for the forest resource needs of the future.

WHY HAVE PRICE AND CONSUMPTION BEHAVIOR CHANGED?

There are at least four hypotheses that might explain the changing consumption and price patterns of the postwar period. Two of these suggest that the decline in the postwar period in the rate of growth of wood demand was accomplished by a moderating of the growth of final consumption of wood and wood products; they thereby explain indirectly the moderation in wood real price rises. The other two hypotheses focus on supply-side phenomena that might result in a moderation of upward pressures on wood prices. The four hypotheses are not mutually exclusive, and anecdotal evidence supports all of them.

The first hypothesis utilizes the observed leveling-off in the developed world of consumer demand for wood products, which occurred in the more recent postwar period in reaction to adjustments in housing stock, the development of nonwood substitutes, and the like. The second hypothesis maintains that wood-saving technologies are moderating the demand for the wood resource for any given level of demand for the final product. Hypothesis three states that past rises in real prices have provided the incentive for forest management and for intensively managed, high-yield forest plantations, thereby allowing supply to expand more rapidly. The final hypothesis is that wood-extending technologies are making more of the world's physical wood base economically available and thereby allowing the effective supply to expand (shift) more rapidly than might have been expected.

The first hypothesis focuses on demand factors and maintains that fundamental changes are tending to moderate the growth of demand over time. The early postwar period was one of rapid economic growth. The destruction wreaked by the war, together with the dearth of construction during the worldwide depression of the 1930s, resulted in a relatively small stock of aging housing and other wood-using structures in the industrial world in the early postwar period. The postwar economic boom, which swept the industrial world, brought on a period of prosperity for the forest products industry. The boom in housing construction extended into the 1970s and continued into the 1980s as a result of the postwar "baby boom," which provided a large new generation requiring additional housing. The baby boom was not unique to the United States; it was also experienced to a lesser degree in both Europe and Japan.

Today, however, residential construction has largely caught up with the market throughout the industrial world. The United States finds itself in the final stages of adjusting its housing stock to the last of the baby boomers, who are now in their late twenties. Even the most optimistic forecasters anticipate that demographics will generate lower housing construction in the 1990s.

The situation on the wood fiber side is similar in the sense that the postwar period may be viewed as the time during which demand growth declined as the industry matured. Using an S-curve type of growth analysis, one might expect to see the industry develop slowly in its infancy, grow more rapidly during its adolescence, and finally experience a declining rate of growth as it moves into its mature phase. This view would be consistent with the reduction in the growth of consumption of pulpwood that has been experienced worldwide over the 1950–1985 period.

The second hypothesis also focuses on the demand side and maintains that wood-saving technology is resulting in continuing demand for the resource to grow less rapidly than the demand for the final product. An important factor to take into consideration in any analysis of future demand for industrial wood is technology and likely future technological changes (see chapter 6). In this regard, several trends on the pulp and paper side are worth consideration. For example, there is a trend toward the increased production of chemi-thermal-mechanical pulp (CTMP) and other wood-saving technologies. CTMP is a wood-saving process that combines some of the features of the chemical and mechanical pulp-making techniques. In the process about twice the volume (by weight) of wood pulp can be produced from a given volume of wood as is realized from traditional processes. Hence only a little over 2 m^3 of pulpwood are required to produce a metric ton of wood pulp, whereas with conventional sulfate processes over 4 m^3 are required per ton.

Although CTMP is not substitutable for traditional pulp in all uses, it is an adequate substitute in many uses, and the applications are being expanded. The implication of this technique is that for a given demand for the final product, the derived demand for the raw material wood input will be reduced. Thus, even if final demand should continue to expand at historical rates into the indefinite future, the introduction of new wood-saving techniques will result in a gradual reduction in the growth of demand for pulpwood.

A third hypothesis is that the rising real price of industrial wood has generated additional sources of supply. A price situation exists in many regions such that it now makes economic sense to plant, manage, and harvest trees, that is, to establish intensively managed forest plantations. The movement to high-yield forestry has been facilitated by an improved understanding of the biology of the forest and by improved and lower-cost techniques and methods for implementing high-yield forestry. Today, plantation forests have been established in many areas that have not traditionally been major sources of industrial wood (for example, New Zealand and Brazil). In addition, traditional producing regions that are experiencing major harvesting and that have favorable economic and biological conditions are being intensively reforested after harvest (for example, in the U.S. South).

Another new source of forest production induced by the increased economic incentive created by higher real wood prices is that of previously inaccessible sites of natural old growth forest now made economically accessible because of the higher prices and/or lower costs of access generated by technical improvements in logging and transportation. Higher real prices can thus facilitate increased output from several sources.

The fourth hypothesis is that improving technology allows previously nonmerchantable wood to be substituted in many production processes. In essence, technology is "wood extending." The increase in the United States in the use of hardwood for pulp production from 14 percent of total wood input in 1950 to 38 percent in 1986 demonstrates the long-term trend toward the substitution of short-fiber hardwood for traditional long-fiber conifer (USDA, Forest Service, 1988, p. 15). Although such a shift may not reduce the total wood requirements for pulp products, technologies of this type allow for the substitution of previously uneconomic low-quality wood for the more traditional wood inputs. Examples of these types of technologies include press-drying techniques as well as new pulping technologies that allow a dramatic increase in the proportion of hardwood fibers. Hence, by increasing the available usable wood, such technologies are supply enhancing. Also, the availability of these inexpensive substitute woods places an effective cap on wood prices.

A similar phenomenon is occurring with solidwood, where new technology is allowing previously inferior and uneconomic roundwood to be substituted for the higher-quality sawtimber resource in the production of solidwood products. One of the outstanding examples of this phenomenon is the development of waferboard and oriented strand board (OSB), which have utilized the "low-quality" aspen resources of the Lake States and Canada to produce a product that directly competes with plywood in many uses. The new technology uses low-quality wood. The lower price of this product and its wood inputs places a cap on the prices of plywood and on the wood resource used as an input. The effect is to relieve the constraints on the economic supply of wood and extend the basic wood resources of the economy.

The first two of the foregoing hypotheses provide an explanation of the observed decline in the rate of growth of worldwide demand for industrial wood. The third hypothesis provides a supply-side response to higher prices that is facilitated, over time, by technological improvements. The fourth hypothesis provides an explanation of how cost-reducing technological improvements could have facilitated a rapid expansion in economic supply, thereby offsetting the effects on wood prices of modestly rising demand.

INTENSIVE FORESTRY, EMERGING PLANTATIONS, AND BACKSTOP TECHNOLOGY

These four hypotheses do not address directly the question of why real prices of most industrial woods have moderated their rise in the post-1950 period. An explanation of the observed behavior of a long-term real price rise is given by several conceptual considerations presented below, which are related to the dynamics of the forest drawdown and the role of technology, most of which are formally embodied in the TSM.

In the context of an old growth forest such as that found in North America by the settlers, the resource is gradually drawn down for the timber values. The real price rise will follow the time path laid out by Hotelling (1931) and modified by Lyon (1981) to reflect the renewability features of the forest. Gradually, the prices of the old growth will rise as it becomes less accessible and abundant. As the real price rises and the existing inventory declines, it becomes economically rational gradually to conserve the existing forest for future harvests at higher prices and also to undertake investments in forest management and tree growing as an economic proposition justified by anticipated future economic returns. As the system moves from reliance on old growth to wide-scale use of managed and plantation forests, the tendency for the real price to rise will moderate.

If technological change is introduced into the system, the real price rise can be further moderated and even reversed. A discussion of this issue is found in Robert Solow's Presidential Address to the American Economic Association (1974). In that address Solow notes that Hotelling (1931) had pointed out the fundamental principle that "if we observe the market for an exhaustible resource at equilibrium we should see the net price . . . rising exponentially." As noted earlier, although industrial wood is not an exhaustible resource, Lyon has shown that a similar outcome will apply to renewable resources. Solow points out that the seriousness of the resource exhaustion problem must depend on the likelihood of natural-resource-saving technological progress and on the ease with which other factors of production can be substituted in production for the resources. These two considerations—resource-saving technological progress and substitution in production—are hypothesized for wood products. These considerations set the stage for the concept of a "backstop technology"—that is, a situation in which at some finite cost production can be freed altogether from the constraints imposed by the nonrenewable aspects of the resource.

In this context the postwar stabilization of long-term real wood prices is indicative of the "balancing" of growth in demand for the wood resource with supply-side responsiveness as assisted by changing technology. It should be noted that this need not imply a necessary reduction in the growth of demand for the final product, but rather a reduction in demand for the industrial wood input as a result of wood-saving technology and/or an increase in effective supply as technologies emerge to extend the economic wood supply. The postwar real price evidence indicates that movement toward a long-term balancing has occurred. Although this "balance" may have been interrupted over any particular period or decade, the post-1950 period experienced overall a long-term balancing of industrial wood for solidwood products as reflected in the absence of strong continuing real price rises.

For solidwood an adequate backstop technology seems to have allowed for the gradual use of smaller and smaller logs, and recently developed techniques have allowed wood chips from previously undesirable species to be used in the production of panels that are substitutes for plywood and lumber in an increasing number of applications. These production technologies, together with improvements in forest management technologies, have allowed the wood resource from intensively managed, rapidly growing plantations to be increasingly substituted for wood from naturally occurring forests. In the context of these technological improvements, plantation-grown timber has emerged as a "backstop" technology. The favorable real price rises that had preceded the advent of the forest plantation provided sufficient financial returns to justify tree growing as an economic activity.

For pulpwood, as noted, the real price data suggest that there never was a period when the resource experienced increasing economic scarcity. However, technological change undoubtedly played an important role in the stabilization of the real price since the large-scale utilization of wood fiber in papermaking at the beginning of the twentieth century. Stabilization occurred with the development of a host of wood-extending and wood-saving pulpmaking techniques. New techniques were continually developed that also effectively expanded economic supply as more species and different wood fiber types were utilized economically. These included pine used for newsprint, the increased substitution of short fiber for long fiber, and so forth.

The diminishing role of old growth forests as a source of wood supply from more accessible managed forests, intensively managed forest plantations as substitutes for inaccessible old growth sources, and the effects of technology on both timber processing and timber growing undoubtedly have played a crucial role in moderating the rise in real wood prices for solidwood resources that has been observed worldwide since 1950, and in maintaining constant real prices for pulpwood since the turn of the century. These supply-side factors, together with the overall decline in the growth rate of demand resulting from both demographic considerations and the role of wood-saving technology, work together to provide an integrated explanation of historical trends in wood prices and consumption patterns.

TOWARD A FORMAL TIMBER SUPPLY MODEL

The Timber Supply Model developed in the following chapters explicitly and formally incorporates several factors needed to explain the historical path of real prices and economic scarcity that have been discussed in this chapter. By incorporating varying degrees of intensity of management, the phenomenon of emerging forest plantations, and wood-saving and wood-growing technological changes, the TSM captures important historical and contemporary factors that have affected timber supply and that explain past price trends. As incorporated into the TSM and this study, these factors can help anticipate future price trends.

REFERENCES

Bourke, I. J. 1986. "Trade in Forest Products: A Study for the Barriers Faced by the Developing Countries," report prepared for the Food and Agriculture Organization of the United Nations (Rome, FAO).

Clawson, Marion. 1979. "Forests in the Long Sweep of American History," *Science* vol. 204, no. 4398 (June) pp. 1168–174.

Economic Commission for Europe/Food and Agriculture Organization of the United Nations (ECE/FAO). 1986. *European Timber Trends and Prospects to the Year 2000 and Beyond*, vol. 1 (New York, United Nations).

Food and Agriculture Organization of the United Nations (FAO). 1983. "Forest Products Prices: 1963–1982," FAO Forestry Paper no. 46 (Rome, FAO).

Hotelling, H. 1931. "The Economics of Exhaustible Resources," *Journal of Political Economy* vol. 39 (April) pp. 137–175.

Lyon, Kenneth S. 1981. "Mining of the Forest and the Time Path of the Price of

Timber," *Journal of Environmental Economics and Management* vol. 8, no. 4, pp. 330–344.

Manthy, Robert S. 1978. *Natural Resource Commodities—A Century of Statistics* (Baltimore, Md., The Johns Hopkins University Press for Resources for the Future).

Milliken, Russell B., and Frederick W. Cubbage. 1985. "Trends in Southern Pine Timber Price Appreciation and Timberland Investment Returns, 1955 to 1983," Research Report 475 (Athens, University of Georgia, College of Agriculture Experiment Station).

Solow, Robert. 1974. "The Economics of Resources or the Resources of Economics," *American Economic Review* vol. 64, no. 2, pp. 1–14.

U.S. Department of Agriculture (USDA), Forest Service. 1988. "U.S. Timber Production, Trade, Consumption, and Price Statistics 1950–1985," miscellaneous publication no. 1460 (Washington, D.C., USDA).

4

The Model and Its Assumptions

This chapter presents a nontechnical discussion of the Timber Supply Model, its characteristics, and some of its important assumptions. The model is designed to address some of the questions raised in chapter 3 using the concepts introduced briefly in that chapter. Together with chapter 5 (on demand) and chapter 6 (on technology), these three chapters provide the essential features of this study's analytical approach. (A highly technical presentation of the optimal control TSM is developed in chapter 7 and amplified in appendix O.)

THE PURPOSE OF THE MODEL

Within the context of this study, the purpose of the model is to function as a tool to assist in the task of assessing the adequacy of the long-term world timber supply. The model is a useful vehicle for systematizing and formalizing the factors that affect long-term supply as well as the nature of the forces and interrelationships that exist within and among supplying regions.

The ability of models to examine the implications of alternative assumptions and situations is one of their major strengths. A formal model allows the user to examine possible futures under various assumptions regarding relationships and events. It also forces the modeler and user to confront the implications of the assumptions and posited structures that affect the projections. Fundamental deficiencies in a model's structure and underlying assumptions or deficiencies in a modeler's analytical process can be identified in the form of implausible projections.

For example, a model that posits demand growing at some predetermined rate for a very long period is likely to overwhelm any supply function that lacks an argument to capture the effects of the growth of technology. This lack will be manifested in dramatically rising real prices. Are such prices realistic, or are they an artifact of an unrealistic assumption or model structure? Such a question raises further questions regarding the usefulness and realism of a model in which demand continues to grow over time while the long-term supply process ignores technological change.

Although analysis rather than forecasting per se is the principal use of models, the process of analyzing various scenarios and assumptions may lead to a most-likely case, which in turn may be designated as a forecast. In this study our base case is designated as a forecast in the sense that we view the projections of the base case as our best "point estimate" of the "most likely" intertemporal time path of world prices and aggregate industrial wood production. Implicit in that interpretation is the judgment that the assumptions of the base case are the most likely to persist over time.

THE CONCEPTUAL APPROACH

One feature of the forest resource that makes many issues in the economics of forestry more complex than they are for many other commodities is the supply curve of the forest resource (see chapter 7). In the short run one can view the economic supply of the forest resource as similar to that of an exhaustible resource. The short-run supply curve relates to the existing stock of forest inventory, the marginal costs of extraction, and the costs of transportation to markets. However, focusing on these factors ignores the interesting question of the approach to the long-term equilibrium. Forest resources, as we have seen, are renewable. Long-term timber supply relates to the rate of drawdown of the existing inventory and also to the rate of renewability of the forest, that is, the change in the existing stock of the forest inventory. The rate of renewability is in turn a function of the level of investment in forest regeneration as well as the natural endowments of the site. Hence, although the problem in the short run deals largely with the economics of timber extraction, the question of long-term supply relates importantly to the costs of timber farming—that is, the costs of planting, growing, and harvesting trees as a crop.

A basic question that the model is designed to address is that of determining the economically optimal transition from an old growth forest to a "regulated" steady-state forest. At what rate should society proceed in the process of drawing down the old growth inventory and replacing it with regenerated forest? Furthermore, what is the appropriate (optimal) level of regeneration investment in the regenerated forests?

The usual way of modeling forest harvest is the simple "growth-drain" approach, in which supply is made a function of the inventory or stock of forest biomass. The Forest Service's Timber Assessment Market Model (TAMM) is a variation of this approach. Because the growth-drain approach has no provision for the age distribution of the inventory, harvests are invariant to the age composition of the forest. In such an approach there is no provision or mechanism for a transition to an even-aged regulated forest and no movement toward a long-term equilibrium. Implicit in growth-drain, therefore, is an assumption that the existing stock of timber is optimal and thus that the initial stock approximates the efficient steady-state situation. For a world in which the forests are in transition from old growth to steady-state harvests, such an approach clearly has serious limitations.

Although the growth-drain approach implicitly assumes no foresight on the part of the harvester, the TSM is a rational-expectations model that assumes that harvesting decisions are on average correct regarding their expectations of future harvest levels and prices.

Conceptual Origins

The conceptual approach of this study draws heavily on both the forest economics and resource economics literature. The concept of discounted present value in forestry goes back to Faustmann (1849), and the classical analysis of the economics of a nonrenewable resource was developed by Hotelling (1931). In Hotelling's context the problem is one of determining the optimum rate of drawdown of the stock of a nonrenewable resource. Walker (1971) developed a "timber harvest scheduling model" that examined the question of the optimal rate of drawdown of an old growth forest and generated an initial time path similar in many respects to the TSM (Lyon and Sedjo, 1986). In recent years the economics of the steady-state forest was developed by forest economists such as Vaux (1973) and Hyde (1980).

These two alternative approaches—old growth stock drawdown and steady-state forestry—were unified by Lyon (1981), who extended Hotelling's nonrenewable model to a renewable resource such as timber. This approach was then expanded and refined by Lyon and Sedjo (1983). In addition, the model as developed allows for the inclusion of features similar to that of Solow's (1974) "backdrop" technology.

The TSM is expressly designed to incorporate into a single framework the elements necessary for examining the question of the transition from a "mining mode" of exploitation of what is treated as an essentially nonrenewable, old growth forest resource to continuous management of a renewable resource.

A Control Theory Approach

The TSM utilizes a control theory approach that explicitly introduces initial conditions and laws of motion for the forest system and control variables monitoring and describing the changing "state" of the forest. The initial conditions refer to factors such as area of forest by age group and land class. The laws of motion describe rules that govern the system over time. For example, over time young trees become old, old trees are harvested, and investments in regeneration are made that in turn influence the rate of growth of the forest. The control variables, such as rotation ages (harvest levels) and the magnitude of regeneration input each year by land class, in turn affect the "state" of the system.

In a control theory approach such as that used by the TSM, the changing age and volume conditions of the forest are constantly monitored and updated so that management decisions can recognize explicitly the changing state of the forest. For example, harvest and regeneration investment decisions depend in part on the state of the existing inventory. The control theory approach adds a degree of realism lacking in the common growth-drain approaches that simply make harvest a function of gross inventory independent of any age considerations.

A General Versus Partial Equilibrium Approach

The TSM represents a partial equilibrium approach in that the forestry sector is modeled and manipulated on the assumption that the nonforestry sectors of the economy are not substantially disturbed by changes in the forestry sector. This sector is linked to the general equilibrium economy through the interest rate, the prices of the various factors and intermediate inputs, and demand for the forest resource. However, the forestry sector is viewed as a price taker with respect to the interest rate and the prices of all the various intermediate and factor inputs utilized in timber production (except forestland). Also, the demand for the forest resource is assumed to be independent of supply-side manipulations. Finally, the TSM is not a trade model and does not project interregional trade flows. Rather, the model projects regional supply and global demand. Regional supply is linked to global demand, and a factor covering the transport cost to major world markets is incorporated (appendix I). However, global demand is not disaggregated by region; hence regional demand is not provided, and interregional trade flows are not estimated.

AN OVERVIEW OF THE MODEL

A working hypothesis of this study is that in the aggregate, timber production in the real world is experiencing a transition from the drawdown of existing old growth stands to utilization of second-growth and plantation-grown industrial wood. This transition, though global, is at different stages in different regions. These regions are treated as subsets of the overall global forest.

One implication of this transition is the possibility of an erosion over time in regional comparative advantages in timber production, as regions that happen to be well-endowed with old growth may be poorly suited for rapid tree growth and other conditions normally associated with high-yield forest plantations (Sedjo and Lyon, 1983). Therefore, an abundance of old growth inventory may ensure a regional comparative advantage only until that inventory is depleted. A continuing advantage would then depend on favorable timber-growing conditions that the region may or may not possess.

The Natural System

The model incorporates physical and biological elements to provide a natural system framework or what economists might call an underlying biological production function. Seven major producing regions are modeled. Each region is disaggregated into a relatively few homogeneous land classes—a total of twenty-two across the seven regions—which are the basic units of production (see appendix A). For each of the twenty-two land classes the model incorporates physical and biological information to develop a production function (appendixes B through H). This includes information on land class quality, location, accessibility, and area; growth yield functions by dominant species and land class; existing inventories and their age distribution; suitability of timber for sawlogs or pulping; and silvicultural responses to investment inputs (appendixes N and M).

Cost Information

In addition to physical information, the model incorporates cost information, including access costs, logging and mill transport costs, and regeneration costs. In recognition of the importance of location vis-à-vis major world markets as a factor determining regional supply, the model builds international transportation costs between each supply region and its principal world market into the cost structure that underlies each regional supply function (appendix I). Finally, the model introduces an economy-wide interest rate that reflects the opportunity cost of capital (appendix K). Because the major purpose of the model is to examine in detail supply-side responsiveness to changing market conditions, most of the detail and data of the model are directed at supply-side considerations.

Economic Optimization

Imposed on the natural system framework is an economic optimization model. This model is designed to project total world timber harvest as the aggregate of the harvests of the major producing regions of the world. Each region is optimized within the context of the overall global system as noted, since the major purpose of the optimization model is to examine economic supply; consequently, the incorporation of supply-side production relationships and data into the model structure is emphasized.

A single exogenously determined world demand function (see chapter 5), adjusted for locational factors as discussed earlier, is imposed on the model and interacts with the aggregate supply produced by the seven separate supply regions. Although this does not allow for the development of international trade flows, it does provide projections of intertemporal harvest levels by major region as well as global estimates of intertemporal harvests and prices.

Any assessment of the long-term timber situation requires that attention be given to demand as well as to supply-side considerations, and the assessment of demand has received considerable attention in this study. The details of the specification of the level and growth rate of demand used in the model are discussed in chapter 5.

Features of the Solution

For the exogenously determined level and growth rate of demand, the model simultaneously solves for the single world market price and the economically optimizing harvest levels for each of the various regions by relating the cost and production conditions of each of the regions to the exogenously determined world demand for industrial wood. The harvest levels of the individual regions are the aggregate of the land classes that make up that region, and the total world harvest is simply the aggregate of the harvests of the individual regions. In addition, part of the solution process is a determination of the economically optimizing level of investment in regeneration by land class.

A particular land class need not enter into the solution or may enter into the solution only under certain circumstances. Thus, for example, an inaccessible region with high harvest costs may have no harvests if a weak demand generates a low price but may become an important supply source if a strong demand generates high prices. Similarly, a region with a long rotation and a very skewed age distribution of trees may be a major producer of timber when most of the trees are of rotation age but may be inactive for long periods while the forest moves toward a mature age distribution. The extent to which a land class moves in and out of production is constrained to some degree by the thirty-year life assumed for all wood-processing facilities. This assumption captures the real-world lack of fungibility of capital.

A unique feature and particular strength of the TSM is its ability to estimate endogenously the economically optimal amount of investment in forest regeneration by land class under varying economic conditions. Utilizing information on biological responsiveness to the level of silvicultural inputs, as well as the expectations of future prices as provided by the simultaneous solution of the model, the TSM solves for a unique level of investment in regeneration for each land class. Consistent with most studies of investment in regeneration, the model results suggest that the better the biological responsiveness of the site, the greater the optimal amount of investment for that site. Thus, after adjusting for economic factors, the model invariably projects higher levels of regeneration inputs for sites classified as high than for lower-class sites.

Casual observation tends to support this result, as it appears that most of the investment in regeneration by the private sector is made in well-located, high-productivity sites. The often-discussed shift of the U.S. forest products industry from the West to the South can be explained in terms of this consideration.

CHANGING OUTPUT LEVELS

In the TSM harvest levels can be affected by adjustments of six types:

1. rotation lengths,
2. the rate of drawdown of old growth inventories,
3. the number of forested land classes utilized for harvest,
4. the level of regeneration input,
5. the rate of biotechnological change in timber growing,
6. the rate at which new industrial plantations are added to the world's forest-producing regions.

The first three of these relate to the extent to which and rate at which existing forests are harvested. These three means of affecting the harvest level might be viewed as short- or medium-term effects. The last three means of affecting harvest levels relate to the rate at which essentially new sources of industrial wood are made available or used more efficiently. The first five of these output adjustments are determined within the model (endogenously) and are affected by the current and future price as well as the interest rate. The sixth —the rate at which new industrial plantations are established and the rate at which technological change affects the rate of growth of timber or the efficiency with which timber is processed—is determined outside the model (exogenously) on the basis of past experience.

In the shorter term the intertemporal levels of harvests can be modified to become more efficient by altering the rate of drawdown and the rotation length in response to economic signals. Also, higher real prices allow marginally accessible forest inventories to be harvested, as the higher prices now justify the higher harvesting and transport costs. Both the harvest of old growth and the modification of the rotation length allow for the replacement of older, less rapidly growing stands by younger, more dynamic stands and thereby change the total volume available for intertemporal distribution. However, in the longer term it is largely the levels of investment in regeneration (or lack thereof), the rate of new plantation establishment, and the effects of technological change that dominate the determination of harvest levels.

The control theory model adapts to the dynamic nature of the forest resource by postulating laws of motion for the system that are necessary to reach a solution. Trees become older and larger if they are not harvested. If they are harvested, a decision is made as to whether to regenerate and, if so, as to the level of regeneration investment. Subsequent growth rates depend on the level of such investment. Within this system the solution procedure identifies the economically optimal time path of price and harvest by solving first for the steady-state optimal levels of harvest by land class and the world market price. The procedure then chooses the optimal or economically most efficient time path—that is, the one that maximizes the sum of consumer and producer surpluses—from among the host of feasible paths between the initial situation and the steady state.

INTERREGIONAL RELATIONSHIPS

Of interest in our examination of long-term harvest levels are not only global potentials but also regional possibilities. In an integrated global system such as currently exists for wood and wood products, it is useful to think in terms of a worldwide price for industrial wood. In fact, of course, there is a spectrum of wood prices; nevertheless, these prices are not independent and reflect the underlying economics of quality differences, location, and transport costs. Thus, the price within each region is determined not only by conditions within that region but also by systemwide conditions; hence the harvest of a particular region depends in large part on the conditions of the other regions. Finally, because there are typically a large number of independent timber decision makers in our regions, each economic agent is assumed to act as a price taker; that is, each agent views that price as determined independently of any actions or decisions that the agent might make. Therefore, although the price of timber and its time path are determined endogenously within the model, the model assumes that producers in each region view the price as external and independent of their particular actions.

In the face of this externally determined price, producers in each region make largely independent decisions, first to harvest and then about whether and to what extent they should invest in forest regeneration. At first glance the harvest decision can be viewed as essentially akin to a mining decision: do the economics justify extracting the existing timber, in light of the present cost and price structure and expectations of future prices? For some set of current and future prices in our model, the answer for some of the twenty-two high-harvest-cost land classes is no; and those land classes do not enter into the solution. Most of the twenty-two land classes in our model are being harvested most of the time.

The harvest decision has an additional component over that of the simpler mining decision when one considers the renewable nature of the forest resource and the opportunity costs associated with maintaining a mature, slowly growing forest when it might be replaced by a young, more vigorously growing forest. In essence a high-productivity site will enhance the returns to regeneration and thus speed up the harvest process. Once the forest has been harvested, the regeneration decision is in principle a second, independent economic decision. In practice, of course, the state often imposes a reforestation requirement. Nevertheless, the requirement typically necessitates only a minimal effort toward reforestation, and the decision must be made as to whether investments in regeneration should exceed the minimal requirements. In this model this is an economically optimized decision based on the incremental costs of regeneration and the incremental returns adjusted for the discount rate.

Over the long time periods examined in this model, old growth will decline as a source of industrial wood, and investments in regeneration will have an increasingly important influence on the extent to which (1) regions and land classes become or continue to be important producers of timber and (2) land classes tend to find their role as producers declining. In the long term, profit-seeking investments will tend to move to regions with a comparative advantage in timber growing and away from low-return regions.

SOME DETAILS OF THE MODEL

The timber-supplying regions of the world can be grouped into ten regions. The first seven are characterized as responsive:

1. U.S. South,
2. U.S. Pacific Northwest,
3. Western Canada (British Columbia),
4. Eastern Canada,
5. Nordic Europe,
6. Asia-Pacific, and
7. the emerging region.

The last three are termed nonresponsive:

8. the Soviet Union,
9. Europe (not included in responsive regions), and
10. all other regions.

The first seven regions are defined as responsive because they are assumed to respond to profit-maximizing incentives and therefore generally behave in a manner roughly approximating what we have termed as economically optimizing. This characterization seems reasonable, as these regions have predominantly market economies and, despite some well-known constraints on harvests in some subregions, generally have an active private forest industry that is functioning largely in response to market forces. These seven regions are subdivided into twenty-two relatively homogeneous land classes. Each of the twenty-two land classes is then explicitly modeled along the lines discussed earlier to capture both its physical and economic conditions.

The names of the seven responsive regions are largely self-explanatory except for the so-called emerging region. This region is defined as an aggregate of tropical and semitropical subregions that have established forest plantations of nonindigenous species. The emerging region includes Brazil, Chile, Venezuela, New Zealand, Australia, South Africa, Spain, and Portugal.

The last three regions—the Soviet Union, other European subregions, and all other regions—are characterized as nonresponsive; their harvest levels are viewed as autonomous and determined independently of the usual economic calculus. This characterization is clearly only an approximation; some European and "other" regions no doubt respond to market incentives, and some areas of the responsive regions surely do not. However, treating the last three regions as a nonresponsive group appears to best approximate the likely behavior of the group in the aggregate.

There are several reasons for treating these three regions as nonresponsive. First, because the Soviet Union and the Eastern Bloc nations of Europe are centrally planned economies, their economic decisions are not expected to be particularly responsive to the types of economic costs and incentives built in to the model. Consequently, no serious thought was given to trying to incorporate these countries' forests into the formal modeling, and their effect on the system is manifested exogenously through assumptions and scenarios regarding the future production of the nonresponsive region. For Western Europe, which consists of basically market economies, the exclusion rationale is simply that many of the forests of Western Europe are managed using criteria different from those embodied in the model. For example, the very long rotations in parts of Europe appear to reflect a preference for rotations longer than those which are financially optimal as well as inheritance tax systems that allow intergenerational transfers of wealth through forest ownership with a minimum of taxes and provide a vehicle for avoiding large inheritance taxes. Furthermore (and somewhat surprisingly), data for much of Western Europe on the age distribution of the existing inventory are unavailable.

Although the TSM recognizes that most regions produce both sawlogs and wood fiber, no attempt is made to differentiate between the two outputs for a given land class. One rationale is that the two forms of the wood resource will become increasingly substitutable for each other as technological change advances. Nevertheless, because the rotation lengths of sawlogs and pulpwood tend to differ, reflecting the value associated with size for sawlogs, the model does characterize each of the twenty-two land classes as producing either sawlogs or pulpwood. This characterization is reflected in different rotation lengths for sawlog and pulpwood forests and thus influences intertemporal harvest levels.

The projections and analyses of the model are driven by the responsive regions. The nonresponsive regions are posited to expand production initially at 0.5 percent annually, falling to zero after fifty years. This modest amount of posited expansion is in keeping with the expansion of these regions over the past three decades. The assumption of the model is that the adjustment to market forces for timber will occur primarily in the responsive sectors.

REFERENCES

Faustmann, Martin. 1849. "On the Determination of the Value Which Forest Land and Immature Stands Possess for Forestry," *Allgemeine Forst und Jagd-Zeitung* vol. 25, pp. 441–445.

Hotelling, H. 1931. "The Economics of Exhaustible Resources," *Journal of Political Economics* vol. 39 (April) pp. 137–175.

Hyde, William. 1980. *Timber Supply, Land Allocation, and Economic Efficiency* (Washington, D.C., The Johns Hopkins University Press for Resources for the Future).

Lyon, Kenneth S. 1981. "Mining of the Forest and the Time Path of the Price of Timber," *Journal of Environmental Economics and Management* vol. 8, no. 4, pp. 330–344.

Lyon, Kenneth S., and Roger A. Sedjo. 1983. "An Optimal Control Theory Model to Estimate the Regional Long-Term Supply of Timber," *Forest Science* vol. 29, no. 4, pp. 798–812.

———. 1986. "Binary-Search SPOC: An Optimal Control Theory Revision of ECHO," *Forest Science* vol. 32, no. 3, pp. 576–584.

Sedjo, Roger A., and Kenneth S. Lyon. 1983. "Long-Term Forest Resources Trade, Global Timber Supply, and Intertemporal Comparative Advantage," *American Journal of Agricultural Economics* vol. 65, no. 5, pp. 1010–1016.

Solow, Robert M. 1974. "The Economics of Resources or the Resources of Economics," *American Economic Review* vol. 64, no. 2, pp. 1–14.

Vaux, Henry J. 1973. "How Much Land Do We Need for Timber Growing?" *Journal of Forestry* no. 70, pp. 399–403.

Walker, John. 1971. "An Economic Model for Optimizing the Rate of Timber Harvesting," Ph.D. dissertation (Seattle, University of Washington).

5

Introducing Demand in the Timber Supply Model

This chapter expands and largely completes the elements necessary for the development of the Timber Supply Model (chapter 4) and anticipates the discussion of technology (chapter 6). We begin by briefly discussing the reason for introducing a timber demand schedule into the model, together with the simultaneous demand-supply nature of the system. Next we examine the effect of various assumptions of demand growth on the projections generated by the TSM. We then present the formal nature and precise specification of demand used for the base case of the TSM. Finally, we present other projections of the future growth of timber demand to allow for comparison with the demand assumptions used in the base case.

Any analysis of future harvests and prices is going to be determined simultaneously by both supply and demand. Sensible projections thus require a reasonable assumption regarding the rate of growth in demand as well as a well-developed model of supply. In this analysis an initial or base-case demand function is specified along with a rate of outward shift of that demand curve through time. The base-case rate of shift of the demand curve is the outgrowth of the quantitative and qualitative analysis of chapter 3. Subsequently, alternative assumptions regarding the growth of demand are made and the implications of alternative rates of growth are investigated via scenario analysis.

The base level of demand and growth rate in demand can be established in various ways. For example, the Global Trade Model of the International Institute for Applied Systems Analysis (IIASA) uses the fairly standard approach of forecasting population and income growth with estimates of the per capita income elasticity of demand of the various final wood products

(Kallio and Wibe, 1987). This results in a projection of future demand levels for the final products. A price elasticity by product adjusts the volumes for price effects. From this the derived demand for the wood inputs is determined by using various wood-to-product conversion factors and adjusting these factors for improvements in technology that are anticipated over time. The total industrial wood input requirement is then calculated by summing the wood requirements for each product across all products. Alternative scenarios are developed by using different estimates of population and/or income growth. A different rate of technological change, reflected in different conversion factors, would also imply different levels and growth of derived demand.

In the timber supply study developed here, the level of derived demand—and particularly the growth of that demand over time—is established in a more direct manner. The initial equilibrium level of worldwide production/consumption is set at 1.5 billion cubic meters (m^3) per year, which is the level actually experienced in the mid-1980s. A demand function is specified that is consistent with that equilibrium level. A rate of outward shift in the demand for industrial wood is then specified for the demand function. The rate of shift used for the base case is based on actual post-1950 experience and developed from the analysis of historical price and harvest trends presented in chapter 3. Changes in these prices provide evidence as to the relative growth of demand vis-à-vis the supply availability and intertemporal changes in that relationship.

DEMAND AND SUPPLY AS A SIMULTANEOUS SYSTEM

Conceptual Considerations

Any attempt to assess the long-term supply potential of a system involving a stock of renewable resources requires not only an assessment of the factors that constitute the physical system in which production takes place (i.e., the production function) but also some set of expectations as to the level of future demand or future prices. These expectations provide the context in which investment decisions, which affect the future stock and future production, are made.

In the short and intermediate term in forestry, supply is constrained by the availability of acceptable roundwood, which in turn depends largely on the existing inventory and the size and age distribution of that inventory. If prices should rise dramatically, supply can be expanded by a more intensive harvest of the existing inventory, including harvesting of some younger trees that have not yet quite reached the traditional rotation age; by harvests from more inaccessible areas, because at the higher prices the greater transportation costs can be covered so that previously economically submarginal timberland becomes supramarginal; by utilization of inferior but less expensive species; and the like.

As the time period is extended into the long term, more options for increasing harvest become progressively available. If price expectations are for increases, investments of various types can be undertaken that increase the long-run level of wood production. For example, silvicultural treatments can be undertaken on existing stands, artificial regeneration can be undertaken to increase future timber inventories, and potential harvests and forest plantations can thus be established in regions and on lands that had not previously been important wood producers. Investments in the development of technology can be undertaken to allow increased forest growth, to allow previously unmerchantable wood to be utilized, or to develop wood-saving technology that will allow the scarce input to be conserved. Such investments all affect the level of long-term supply.

For the foregoing reasons any attempt to look at the long-run timber supply must necessarily involve assumptions and assessments of future demand conditions as manifest in future prices. Of course future prices affect both supply and demand simultaneously. Periods of anticipated high wood prices will not only result in investments in additional wood sources that will ultimately expand supply but also create incentives for consumers and competitors to substitute nonwood inputs for wood inputs.

The Effects of Demand on the Projections

Within the context of the TSM with its twenty-two land classes, the effects of demand on future prices and price expectations and their effects on supply are restricted to a limited number of channels. There are five margins in the model for changing the harvest flow:

1. an increase in the drawdown of old growth,
2. a shortening of the rotation,
3. an increase in investment in the intensity of silviculture and regeneration for existing forestland,
4. the introduction of an additional endogenous land class into the harvest solution, and
5. the introduction of biotechnological improvements in production throughout the regeneration process.

All five of these margins will be somewhat responsive to different levels of product price generated by different levels of long-term demand. The first two margins involve an intertemporal modification of harvest flows. For example, to the extent that a higher future demand implies a relatively higher future price, future harvests will compete resources away from current harvests, thereby reducing the current drawdown of old growth harvests. Also, because the optimal economic rotation length is affected by future prices as well as the interest rate, the optimal rotation length will be extended. In addition, the third margin, the level of investment in forest regeneration, is directly related to expected future harvest prices, which again are a function of future demand. Higher future demand elicits increased investment and therefore greater future long-term supply. The fourth margin, that is, the introduction of additional land classes to the economic harvesting land base, is made operative as the higher demand and higher future prices generate sufficient returns to make formerly submarginal timber stands supramarginal. Thus, as in the real world, the higher future demand and prices draw more land classes into the economic timber base and provide for increased harvests. The fifth margin on which the model can adjust, biotechnological change, is introduced via the yield function (see appendix L).

Although many of the important effects of future demand conditions on long-term supply are captured in the model as just demonstrated, several additional real-world influences on long-term supply are not explicitly developed within the formal model. However, some of these effects can be introduced through the use of a scenario approach. For example, the establishment of new net additions to the land area in forest plantations, though not endogenous in the model, has been addressed in the scenario analysis. High, intermediate, and low levels of plantation establishment are explored. Because one would expect that the rate of the establishment of new plantations in the emerging region would be a function of the expectation of future wood price levels, which in turn are related to expectations of future demand, one might also expect that the high-demand scenario would most likely be associated with the scenario of high plantation establishment, whereas the low plantation scenario would most likely be associated with the low-demand scenario.

Another factor that should be introduced into the model is the rate of technological change in wood processing, which allows for the substitution of inferior wood and fibers in processing (referred to here as wood-extending technology). Although technological change is often viewed as autonomous, it is probably more properly treated as "induced" by changes in relative prices, particularly over the long term. That is, increased demand reflected in higher real wood prices is likely to induce wood-extending supply-side technological change. In this study, however, wood-extending technological change is not introduced into the model; hence the estimates of supply are likely to be biased toward more modest levels.

FUTURE CHANGES IN DEMAND

Any projections of future levels of demand for industrial wood are, of course, largely conjectural. Nevertheless, conjectures of this type must be made either implicitly or explicitly by investors interested in timber and timber growing. The foregoing sections have demonstrated that in the postwar period, the rate of growth of worldwide consumption of industrial wood has tended to decrease. It is noteworthy that because the real price of both industrial wood and pulpwood changed very little over the entire thirty-three-year period, the growth in consumption is approximately equivalent to the growth (shift) in the demand curve over that period. In addition, as documented in chapter 3, worldwide demand has tended to shift outward at a decreasing rate through the post-1950 period.

The future is unknown, but in the absence of a strong case to the contrary, it is probably most reasonable to assume that the forces that have resulted in the moderation of demand for industrial wood in recent years are likely to continue in a similar pattern, especially if other arguments tend to support this view. The decade of the 1970s saw a worldwide average annual rate of consumption increase of only 1.1 percent, and the period from 1970 to 1985 saw that average rate drop to below 1.0 percent. Clearly, in recent decades the rate of growth of consumption worldwide has been modest. A simple extrapolation of the postwar trend would show that the rate of growth of demand will continue to decline into the future.

DEMAND AS INTRODUCED IN THE TSM

The complete TSM requires a specification of the demand function. For reasons discussed earlier, it was decided that the future rate of growth or outward shift of the demand function would be estimated exogenously, based on (a) a quantitative assessment of long-run and post-World War II trends, and (b) a qualitative assessment of considerations such as population growth, market saturation, demographics, technological change, and so forth. Both types of considerations have led us to conclude that the demand function for the primary wood resource is likely to shift at rather modest rates in the intermediate and longer-term future.

Because the TSM formally models only about one-half of the current industrial wood-producing capacity of the world, it is necessary to treat the demand that is introduced into the model as an excess or residual demand function. The quantity (volume) of demand from the responsive regions is as follows:

$$Q_j = Q_j^t - S_j^o$$

where Q_j is quantity for the responsive regions, Q_j^t is the total world quantity, and S_j^o is the volume supplied by the nonresponsive regions. The general form of the world demand curve is

$$Q_j^t = a_j + bP_j$$

hence

$$Q_j = a_j - S_j^o + bP_j$$

where a_j is the intercept coefficient in year j, b is the slope coefficient, and P is the wood price. The actual values are

$$\begin{aligned} a_j &= 115 \qquad (t = 0) \\ b &= -0.001215 \\ P &= 115 - 0.001215Q \end{aligned}$$

where a_j is the intercept coefficient in year j and b is the slope coefficient.

The intercept coefficient for the initial year and the slope coefficient were selected as a pair to yield "reasonable" prices and elasticities of demand in the scenarios. The coefficient a_j varies by scenario and year. It is calculated to yield the correct growth rate of world demand for the scenario and year it represents. We calculate the a_j's to yield the correct growth rate at a price of \$30/m^3. We do this because changing a_j yields a horizontal shift of the demand curve, and the growth rate of quantity for a given shift increases as one shifts northwestward along the demand curve (Q decreases but ΔQ is constant). In addition, the S_j^o is calculated to yield the assumed growth rate of the supply of the nonresponsive regions for the scenario and year it represents.

The elasticity of world demand in the base scenario varies from 0.17 in the initial year to 0.18 in the stationary state, and for the high-demand growth scenario it varies from 0.19 to 0.30 for the same two periods. In addition, the elasticity of residual demand (demand facing the responsive regions) in the base scenario varies from 0.35 in the initial year to 0.32 in the stationary state, and for the high-demand growth scenario it varies from 0.37 to 0.49 for the same two periods.

In the base case the assumption is that the worldwide demand grows (that is, functionally shifts outward) at an annual rate of 1.0 percent annually. Because the nonresponsive region's harvest is growing at less than 1 percent annually in the base case, the demand function facing the responsive regions grows initially at about 1.5 percent a year. (For a more detailed discussion, see chapter 8.) The worldwide growth rate used is less than the growth rate of the 1970s but greater than the rate of the 1970–1985 period. Furthermore, the rate of growth of demand in the base case is posited to decline linearly to zero in year fifty. Thus, at the midpoint of the fifty-year period (year 2010) the worldwide growth rate of the base-case demand will have declined to 0.5 percent annually.

The rationale for declining demand is threefold. First, it is reflective of the post-World War II historical experience as presented in table 3-1, in which demand increased at a decreasing rate. In that sense the diminution of the rate of growth of demand is merely a crude extrapolation of an existing trend. Second, as a practical matter we are really concerned only about projections over the next several decades. Longer-term trends begin to lose their meaning as the underlying relationships, even when quite stable, tend to erode severely over very long periods. Thus, the 100-year time period used in the model is required more with a view to facilitating orderly behavior in the model than with a view to the usefulness or relevance of the longer-term projections. Finally, because factors on the supply side, such as the production of the nonresponsive countries and the rate of technological change, are also extrapolated to decline to zero growth in fifty years, such a construction does allow for symmetry between the demand side and the supply side. The bias in such an approach would not be with regard to price and pressures on price; rather, there would be a tendency to underestimate the growth of both demand and production over the period.

As part of the sensitivity analysis, we examine alternative levels of demand growth, both higher and lower, in chapter 8. The low-demand scenario also assumes that wood-saving technology continues to reduce the wood fiber inputs needed for a given volume of pulp and that pulp and paper demand continue to be relatively soft into the indefinite future. The high-demand scenario assumes a growth in global demand that is more consistent with the worldwide experience over the entire period since 1950. This scenario is predicated on the assumption that wood-saving technologies will proceed at only modest rates, whereas the high incomes and preference patterns of certain rapidly growing Third World countries will rekindle a rapid expansion in demand similar to that experienced in the early postwar period.

OTHER VIEWS OF FUTURE DEMAND

Table 5-1 presents several forecasts of the growth of industrial roundwood consumption, which can be compared to the TSM's assumption that the demand for raw wood expands worldwide at a rate of 1.0 percent annually, declining gradually thereafter. As noted, the initial growth rate used for the TSM is slightly in excess of the growth in worldwide consumption for the period from 1970 to 1985 (table 3-1). Because real wood prices were essentially the same at the end of the period as at the beginning (table 3-4), the growth in consumption and the growth in demand were identical. Other forecast growth rates of industrial wood consumption range from 4.2 percent (World Bank, 1979) to 1.2 percent (Food and Agriculture Organization of the United Nations, 1979; Kallio, Dykstra, and Binkley, 1987). Because the World Bank study assumed constant real prices, the necessary implication is that demand was projected to grow at an identical 4.2 percent annually over

Table 5-1. Forecasts of Industrial Roundwood Consumption

Organization/study[a]	Year made	Projected to year:	Forecast volume (billions of m^3)	Implicit growth rate[b] (percent/year)
Food and Agriculture Organization	1982 (high)	2000	2.6	3.7
	(low)	2000	2.3	2.9
Food and Agriculture Organization, Industry Working Party	1979	2000	1.8	1.2
SRI	1979	2000	1.9	1.6
World Bank	1978	2000	2.8[c]	4.2
		2025	5.9[c]	3.4
International Institute for Applied Systems Analysis (IIASA) GTM	1987	2000	1.8	1.2
		2030[d]	2.6	1.2

Source: Adapted from David W. K. Boulter, "Global Supply-Demand Outlook for Industrial Roundwood," *Proceedings of the National Forest Congress,* April 9 (Ottawa, Canadian Forestry Association, 1986) p. 5.

[a] SRI International and World Bank data from Boulter (1986). FAO 1982 data from FAO, "World Forest Products: Demand and Supply 1990 and 2000," FAO Forestry Paper no. 28 (Rome, FAO, 1982). FAO Industry Working Party data from FAO, "FAO World Supply Outlook for Timber Supply," phase V (Rome, FAO, 1979). IIASA data from M. Kallio, D. P. Dykstra, and C. S. Binkley, eds., *The Global Forest Sector: An Analytical Perspective* (New York, Wiley, 1987), various chapters.

[b] From a base of 1.5 billion m^3 in 1985.

[c] Based on constant prices.

[d] The IIASA GTM projections to 2030 are not reported in *The Global Forest Sector: An Analytical Perspective.* They were presented, however, at numerous meetings and in working papers such as "The Global Forest Model" (unpublished draft by Luke Popovich, May 1985).

that period. The FAO and IIASA studies projected consumption to grow at an annual rate of 1.2 percent between 1985 and 2000, an estimated growth only slightly higher than that in this study. However, because in the IIASA study real prices were projected to increase significantly, the implicit annual growth rate of demand was greater than 1.2 percent—perhaps averaging about 1.5 percent.

It should be noted that although the World Bank study substantially exceeds the historical growth rate of consumption of 2 percent experienced in the post-1950 period, the FAO (1979) Industry Working Party study and the IIASA study are approximately midway between the average consumption growth rate of the entire post-1950 period and the 0.94 percent average for the 1970–1985 period.

REFERENCES

Boulter, David W. K. 1986. "Global Supply-Demand Outlook for Industrial Roundwood," *Proceedings of the National Forest Congress,* April 9 (Ottawa, Canadian Forestry Association) p. 5.

Food and Agriculture Organization of the United Nations (FAO). 1979. "FAO World Outlook for Timber Supply." Phase V. Prepared by a Joint Forestry and Industry Working Party for the Forestry Department (Rome, FAO).

———. 1982. "World Forest Products: Demand and Supply 1990 and 2000," FAO Forestry Paper no. 28 (Rome, FAO).

———. 1983. "Forest Products Prices: 1963–1982," FAO Forestry Paper no. 46 (Rome, FAO).

Kallio, M., and S. Wibe. 1987. "Demand Functions for Forest Products," in M. Kallio, D. P. Dykstra, and C. S. Binkley, eds., *The Global Forest Sector: An Analytical Perspective* (New York, Wiley).

Kallio, Markku, Dennis P. Dykstra, and Clark S. Binkley, eds. 1987. *The Global Forest Sector: An Analytical Perspective* (New York, Wiley).

World Bank. 1979. EIS Paper no. 98, prepared by James Gammie (London, IBRD).

6

The Role of Technological Change

Any assessment of the long-run future requires that some assumption be made regarding the rate at which technological change will occur. Models, implicit or explicit, that assume increasing demand through time with stagnant technology and finite resources invariably find growing economic scarcity as reflected by rising real prices. Several decades ago Solow (1956) used a simple model that demonstrated the links between per capita income growth and population growth. In this simple model, growth in per capita income depended on the rate of growth of technological change outrunning the rate of growth of population. In the absence of technological change, the system was condemned to falling per capita incomes. Solow (1957) further clarified the role of such change on aggregate output in his pathbreaking article on this topic. He expanded this analysis in his book on capital theory (1965), which argued that although short-term economic growth may be due in part to the level of savings in the economy, in the long term economic growth depends largely on the rate of technological change.

In the absence of any effects of technological change on supply, most models, either formal or implicit, will invariably be overwhelmed by unceasing growth on the demand side. They will lead to higher real prices when the focus is on a single or small group of commodities, or to declining real incomes and standards of living when the focus is on the aggregate economy. In the real world, at least in recent centuries, many economies have experienced extended periods in which they have had increasing levels of real per capita incomes despite rising populations. This phenomenon was first experienced by a few countries in Western Europe and North America but more recently has spread throughout much of the world, including almost all of

Europe and North America, much of East Asia, and parts of Latin America; most recently the phenomenon has appeared to affect the highly populated countries of India and China. Furthermore, as we noted earlier, Potter and Christy (1962) and later Manthy (1978) have shown that despite the increased pressures placed on natural resources by widespread economic growth and enlarged populations, most natural resources have exhibited no change in economic scarcity as reflected by a horizontal long-term trend in their real market prices.

The most straightforward explanation of what on the surface appears to be a surprising phenomenon, that is, persistent improvements in living standards and stable resource scarcity, is found in the effects of technological change. Simply stated, the supply of most resources has kept up with rising demand, and these resources have not become economically more scarce.

Some resources, of course, have experienced increasing scarcity for certain periods of time. For example, in addition to certain types of industrial wood, energy resources in general and petroleum in particular became economically scarcer during much of the 1970s; this fact was reflected in the rise in their real market prices. However, this phenomenon was relatively short-lived and surely reflected, at least in part, political factors that allowed for the creation of a cartel rather than any discontinuous change in fundamental economic resource availability.

Although there are occasional deviations in long-term trends, the basic fact is that the real prices of most natural resources have exhibited long-term stability. Again, the most straightforward explanation is that the rate of technological change has been rapid enough to offset the expansion of demand. No guarantee exists that this relationship will continue into the future, but there is certainly little evidence that the rate of technological progress will decline. If anything, casual observation suggests that this rate may be increasing. It thus seems reasonable to assume that the existing trend will continue.[1]

BROAD TRENDS IN TECHNOLOGICAL CHANGE AFFECTING WOOD USE

Today a large number of technological changes are under way in the forest products industry. They promise to have significant effects on the industry

[1]One of the weaknesses of the early versions of the IIASA Global Trade Model in projecting forest product production and trade patterns over a fifty-year period was the linkage of technological change to changes in the plant and equipment, which in turn were allowed to change in the model only through the first half of the fifty-year period examined. Thus the ability of the IIASA model to introduce technological change was limited, and the rapid price rises of the latter part of the period reflected this limitation.

and on supply and demand for the forest resource. These changes include both new products and new methods for producing traditional products (U.S. Congress, Office of Technology Assessment, 1984; Papertree Economics, 1986).

New Products

Historically, there have been numerous instances in which new products have displaced wood. Perhaps the most profound example to date has been the substitution of fossil fuel for wood fuel, which has occurred over the past 150 years or so.

A host of different innovations are responsible for this profound transformation. As recently as 1890, about 50 percent of all the wood utilized in the United States was used for energy. This figure fell to as low as 5 percent in about 1970 but rose to an estimated 20 percent in some regions by 1980, in part because of the "energy crisis." It probably declined during the 1980s. Other examples of new products that displaced wood from an important market include the steamship, which dramatically reduced the market for large straight trees for ships' masts, and the advent of a host of modern transportation modes ranging from automobiles to airplanes, which diminished the demand for railroads and thus the demand for wooden railroad ties (see in this regard Olson, 1971).

This process is continuing unabated on a number of fronts. Advances in electronic information technologies raise questions regarding paper usage, particularly in the longer term. Many analysts anticipate that electronic activities will substitute for paper in numerous functions, including video telephone directories, consumer magazines, electronic funds transfer, electronic mail, electronic storage and microforms, home video catalogues, and so forth. Beyond 1995, portable handheld video displays might partially displace printed magazines, books, and newspapers. Although the issue is very debatable, some analysts have maintained that with 30 percent of U.S. paper production being used for the storage and transmission of information, the long-term effect ultimately will be sharp reductions in paper use which could occur over the next fifteen or twenty years.

However, thus far there is little evidence that the effects of electronics have been to reduce total paper use. There surely has been a substitution of electronics for paper, but this effect appears to be overwhelmed, in the short run at least, by the greater overall use of paper in conjunction with the massive overall increase in information flows associated with electronic information systems. An economist might argue that the "output effect" has dominated the "substitution effect" and that the result has been a net increase in the use of paper. In the longer term, however, there is a very real possibility that this relationship may not continue, and paper usage could be adversely affected.

A second area in which new products could affect the demand for forest products is that of packaging materials. Over the years plastics have made significant inroads into certain types of packaging and have displaced paper sacks for light-duty uses. Since the late 1970s plastic expanded dramatically its share of the U.S. small grocery-sack market and merchandise bag market. This latter share was estimated to be 35 percent in 1981, an increase of 10 percent over the previous year. Many analysts expect plastic to dominate this market in a relatively few years.

One reason for plastic's recent inroads into the kraft paper bag and sack market is technological developments in polyethylene that permit much greater strength. Plastic containers for liquids also have made significant gains in displacing glass and coated paper containers. In Europe and much of the rest of the developed world, the plastic sack is in much more common use than it is in the United States, so there is probably less opportunity for paper market erosion outside North America. Although earlier increases in energy costs have improved paper's competitive position relative to that of petroleum-based products, recent declines in petroleum prices are likely to reverse this situation and perhaps encourage a further movement away from paper packaging materials and toward a more rapid substitution of non-wood-based paper substitutions. Of course, over the longer term the price path of petroleum and the affordability of petroleum-based products is conjectural.

A third area in which technology allows for a degree of substitution between wood-based and non-wood-based products is synthetic fibers. Petroleum-based synthetic fibers compete with cellulosic rayon and acetate; however, the cellulosic fibers have been losing market share to the noncellulosic human-made fibers since 1968, when the worldwide volume of the two fibers was about even (3.9 million tons). There was a resurgence in interest in cellulosic fibers while petroleum prices were very high; however, the overall return to these fibers was small, and with the recent decline in petroleum prices a significant reversal of the shift to petroleum-based fibers appears very unlikely.

A discouraging element of the foregoing discussion for the forest industry is the fact that broad trends suggest that forest products are generally struggling to retain traditional markets, not to expand them or move into new markets. There are some exceptions. Innovations that create new wood products with enhanced structural properties suggest the possibility of displacing steel and other metals in some structural uses. Similarly, new techniques for treating wood against fire and decay offer the possibility of expanded wood products in construction, which may displace nonwood products. For the most part, however, innovations have involved the substitution of one wood product or input for another. To the extent that this trend continues, the demand for forest resources will be driven primarily by demands for wood inputs for existing products. This consideration implies that future demand for wood will probably have to come largely from existing wood products or newly developed wood products, which compete with each other and also compete to some extent with new, technologically generated non-wood-using products.

NEW PRODUCTION PROCESSES

The preceding discussion focused on the effects on wood use of technological change that creates new non-wood-using products that compete directly with wood or wood-intensive products. We turn now to certain changes occurring in the production process and explore their implications for future wood demand and supply. As was the case with new products, technological change is occurring in both solidwood production and in pulp and papermaking.

Solidwood Production

On the solidwood side a variety of innovations are occurring in the production process to increase the usage of wood for lumber and plywood production in lumber products and processes. In the early 1980s the lumber recovery factors in the United States averaged about 41 percent; for plywood the factor was around 50 percent. New technologies, some using computerized controls, have been and are being developed to increase the recovery factor. These include such innovative techniques as best opening face (BOF); saw-dry-rip (SDR); and edge, glue, and rip (EGAR), all of which are marginal modifications to conventional sawmilling that increase recovery rates. Moreover, new wood-saving construction techniques are being developed. For example, wood trusses, or pieces of lumber joined together to form framing members, have been widespread in roof framing for decades and are now being used increasingly in the construction of floors and whole house framing.

In plywood, production techniques have been developed that allow the use of smaller logs. The recent development of the powered backup roll (PBR) increases the amount of peelable material, thereby increasing plywood yields. In addition, composite lumber manufacturing can increase recovery efficiency dramatically through the use of waste materials and also hardwoods. The major opportunities here appear to be in larger applications and specialty applications.

Product recovery rates in particleboard, composite panel, and fiberboard mills approach 75 percent. Of particular interest here are the structural composite panels such as oriented strand board, which have structural properties that allow them to compete directly with lumber and plywood in many uses. In 1979 these products accounted for only about 1.5 percent of the structural panel market in the United States; by the mid-1980s, however, this fraction was much higher, and their ability to displace plywood on a massive scale is well recognized. The appeal of these products lies not only in their high recovery rates but also in their utilization of inexpensive, low-quality, abundant hardwood species such as aspen.

Pulp and Paper Products

Substantial wood-saving and wood-extending technologies are also being developed in pulp and paper processing. Production is moving toward higher-yielding pulps that require less wood input per unit of output. A number of new pulping technologies offer considerable promise in this regard. Newly developed mechanical pulping technologies are reducing dramatically the amount of fiber feedstock required to produce a unit of wood pulp. In addition, these technologies are broadening significantly the types of wood fiber that can be utilized as feedstock and especially expanding opportunities to use the plentiful hardwood resource.

The increased utilization of hardwood fibers in pulp production is not a recent phenomenon. Technological changes have allowed the proportion of hardwood fiber in total pulpwood input to rise from about 10 percent in 1950 to 30 percent in the early 1980s (Styan, 1980). In Europe hardwood pulps now account for around 40 percent of bleached pulp usage, up from 24 percent in 1970.

HOW TECHNOLOGICAL CHANGE AFFECTS WOOD SUPPLY AND DEMAND

At the conceptual level there are four basic ways in which technological change can affect the supply of or demand for the roundwood input. Such change can be either (1) wood saving, (2) wood extending, (3) yield enhancing, or (4) harvest-cost reducing.

Wood-saving technological change, as noted earlier, can be viewed as dampening the demand for roundwood because, for a given output of final product, the demand for the roundwood input will be reduced by the introduction of wood-saving technology. Thus, for example, when demand for the final product grows at some rate—say, 2 percent annually—the demand for industrial wood input will grow at a less rapid rate (for example, 1 percent a year), with the difference accounted for by the wood-saving technological change.

Technological change is wood saving if it allows a greater fraction of the biomass from a given harvest to be utilized for production, or if it develops products that require less wood input per unit of wood product output—for example, if the technology increases recovery factors. A change in technology can also be wood extending, that is, increasing supply by allowing a previously inferior resource, not viewed as part of the industrial wood base, to substitute for a higher-quality resource, thereby freeing the higher-quality resource for other uses. For example, if technology allows the previously unused aspen species to be economically utilized (as with oriented strand board), the effect is to increase the economically usable and (typically) the harvested volume of the wood resource. Such a change would be manifested as an increase in the volumes of harvests.

In addition, technological change can be yield enhancing, affecting economic supply through the development of technically more efficient management practices for both natural and human-made forests that increase yields at harvest. An example of such a biotechnological improvement is found in genetically improved seedlings, which allow for the planting of only genetically fast-growing trees and thus enhance yields. Genetic improvement is the same mechanism that has allowed for much of the dramatic increases in foodstuff outputs experienced during this century (Hayami and Ruttan, 1971).

Finally, technological change can take the form of harvest cost-reducing innovations that make inaccessible, submarginal forested areas economical for purposes of industrial wood harvests. This source of technological change is not stressed in this study because, as Williams and Gasson (1986) show, real price rises and real cost declines are symmetrical in their effect on harvest levels. However, a recent study by Cubbage, Stokes, and Granskog (1988) indicates increasing productivity and lower harvesting real costs in the U.S. South. Also, the Williams and Gasson study demonstrates the potential importance of this type of technological change for the economics of logging in old growth regions such as British Columbia.

Some Recent Examples of Changing Techniques

One of the more dramatic examples of a newly developed wood-extending technology is press drying. This technique is likely to facilitate the continuing increase in utilization of short-fiber hardwood species in various fiber uses. In this technique of paper manufacture, the fiber is dried under high pressure and restraint, thus improving the fiber-to-fiber bonding of stiff, short hardwood fibers. This technique makes it possible to increase dramatically the proportion of hardwood fiber in products that were previously largely softwood, such as paperboard.

A major wood-saving technological innovation is found in the pulping process. Ten years ago thermomechanical pulp (TMP) was virtually unknown in any major commercial application. Today it is practically the only accepted way to build new pulping capacity for newsprint, and it is replacing older techniques in existing plants. It now accounts for more than one-third of the pulp used in Canadian newsprint. Because pulp yields from all the mechanical pulping processes are considerably higher than the techniques they replace, the net effect is to conserve the wood resource.

A very promising new pulping technology that is just beginning to have an effect is chemi-thermomechanical pulp (CTMP), which combines elements of the mechanical pulping process with those of the chemical processes. This wood-saving technique almost doubles traditional chemical pulp yields, thereby conserving the wood resource. In addition, CTMP has the advantages of economic-sized mills that are smaller than the traditional kraft mills; the financial commitment associated with a new mill is considerably smaller than that associated with the traditional mill. In the four-year period between 1981 and 1985, the market CTMP capacity increased sixfold to 600,000 tons. In addition, planned additions to pulp capacity show a dramatic increase in CTMP capacity and a corresponding decline in planned new capacity for the chemical pulps (Papertree Economics, 1986).

Some Implications for Forestry

Although pulp is not usable for some grades of paper, changing papermaking technology allows for increased use of material fillers such as clay to provide a product comparable to that using lower-yielding pulping techniques. European papermakers are now using fillers to make up 20 to 30 percent of the sheet weight of certain grades of papers, compared with 10 to 15 percent in the mid-1970s.

In summing up the existing trends in new technology for pulp and paper, Papertree Economics (1986) notes that

> the NorScan countries, and Canada in particular, have enjoyed an extended period of "Softwood kraft is King." That was true twenty years, ten years, maybe even five years ago; but it isn't anymore, and may never be again. Northern softwood kraft has become too expensive, and its exceptional fibre length much less important; it has thus been losing heavily to tropical and European hardwoods.

Although the foregoing cases are examples of new products or new technologies that reduce the demand for the wood resource or enhance the effective supply available to meet demand, in principle the situation could be just the opposite. That is, a new product or technology could be introduced that increases the demand for the wood resource at the expense of other resources. On a small scale this occurred during the energy crunch of the 1970s, when efficient wood-burning stoves replaced electric heating to a modest extent in parts of New England. However, for industrial wood it is difficult to think of examples of a major new wood-using technology that has not displaced other wood-using technologies. One example might be the development of treated wood to be used for certain types of foundation construction. Predominantly, however, new wood-using products or technologies simply substitute new wood products for earlier products made of wood. For example, although the development of waferboard and oriented strand board do represent a new wood-using product developed through technological change, these products are directly competitive with a more mature wood product—plywood. Hence at least in recent times the substitution effect has rarely increased significantly the demand for the basic wood resource.

THE EFFECTS OF YIELD-ENHANCING TECHNOLOGY ON INDUSTRIAL WOOD OUTPUT

As noted earlier, to some extent the characterization of technological change as affecting demand or supply is a matter of methodological convenience. In this study when technological change results in the creation of more industrial wood either by more rapid growth of trees, access to stands not previously economically accessible, or through the use of stands previously deemed nonmerchantable (perhaps because of the species involved), an increase in the economic supply is deemed to occur and would be recorded in the data as an increase in the harvest of industrial roundwood. Where technological change involves greater utilization of wood already harvested or lower wood inputs per unit of product output, the technology is viewed as wood saving and would be reflected in a less rapid outward shift in the demand curve for industrial wood (but not in the demand for processed wood products).

The third major means whereby technology affects output is through the yield-enhancing effects of new technology on tree growth. In recent decades and particularly since about 1960, there is evidence of substantially increased activity in intensively managed forestry and particularly the establishment of industrial forest plantations (Sedjo, 1983). The advent of this activity reflects both a judgment about the adequacy of future timber supplies from naturally regenerated forests as well as the development of an operational technology that makes artificial regeneration and intensively managed, yield-enhancing forestry both technically and economically feasible. As in an earlier time in food production, cropping provides a ready vehicle for introducing technical improvements. The development of tree-planting operations allows for the introduction of genetically superior trees. This, together with intensive management, allows for future increases of industrial wood harvests, sometimes from regions that have never been important producers of industrial wood. As with agricultural cropping, planting management requires decisions about what species to plant and where to establish the plantation forest. Hence the bounds imposed by the natural system no longer apply and forest investments are made, often utilizing exotic species on lands that in some cases have never been forested.

The effect of this situation on the world's forest products markets can be grasped when one realizes that eucalyptus market pulp now accounts for 13 percent of the world's market of bleached kraft pulp, up from 4 percent only eight years ago. Major sources of this pulp include Spain, Portugal, and Brazil. None of these countries was a supplier until recent years, and in none of these countries is eucalyptus an indigenous species.

INTRODUCING TECHNOLOGICAL CHANGE INTO THE TIMBER SUPPLY MODEL

In this study's modeling analysis of long-term timber supply, the influence of technology on tree growing and timber production is formally introduced into the TSM in three different ways. First, as discussed earlier, the wood-saving feature of technological change is introduced through the moderation of the growth rate of the demand curve, reflecting the effects of wood-saving technology. The second channel for introducing technological change into the model is through the establishment of plantation forests in locations that previously were not significant producers of timber. In the model the effect of the plantation forests on supply is captured via the introduction of the so-called emerging region as one of the seven formally modeled regions. In the real world this composite region encompasses several newly emerging wood-producing countries that have established large areas of exotic (nonindigenous) forest plantations of industrial wood. This region has been almost completely created in the past twenty-five years. Its wood production is largely the outgrowth of technological changes that have allowed for low-cost, intensive forest management.

Finally, related to the creation of the emerging region, an additional means of formally incorporating the wood-producing incremental effect of future technological developments is to introduce yield-enhancing technology as exemplified by genetically superior seedlings. These seedlings, introduced as the result of future investments in regeneration, translate into higher growth rates and ultimately into higher harvest volumes.

These mechanisms have been formalized to introduce technological change into the model. However, the wood-extending and harvest-cost-reducing technical changes discussed earlier have not been introduced formally.

THE QUANTITATIVE EFFECT OF TECHNOLOGY ON THE GLOBAL SOFTWOOD TIMBER SUPPLY

Although there might be general agreement that technological change is very important in any long-term assessment of the future supply (and demand) of industrial wood, arriving at a consensus on the rate at which technology is progressing is difficult. In his 1965 study Solow speculated as to the economy-wide rate of technological progress in the United States. He viewed an estimate of 1.8 percent annually as being reasonable. Recently, Capalbo and Vo (1988) estimated the postwar annual rate of technological change in agriculture to be about 1.5 percent.

In the forestry field Newman (1986) estimated the annual rate of productivity change for the growth of the softwood timber resource at around 0.5 percent to 0.8 percent for twelve southern states over the thirty-three-year period ending in 1985. Other studies have estimated the rate of growth of productivity in the wood-processing industry. Robinson (1975) found an annual rate of technological change of 1.75 percent for the wood processing industry over a twenty-one-year period. Risbrudt (1979), examining four wood industry groups, found a large variability in technological change but an annual average rate of 1.9 percent, an estimate certainly consistent with those of other researchers.

In an interesting study Haygreen, Gregersen, Holland, and Stone (1986) examined the effects of eight emerging technologies on the availability of future softwood timber supplies. Their basic approach was to estimate the extent to which a new technology would allow for reduced softwood use through improved lumber yields or the substitution of hardwoods for softwoods in the production process. The wood-savings estimates were made for the final twenty years of the twentieth century. The effect of these technologies would be to provide for reduced pressure on the softwood resource even as the total usable supply was expanded.

The technologies examined for solidwood were the truss-frame system of home construction; the edge, glue, and rip and the saw-dry-rip techniques of improving hardwood lumber so as to allow for its greater substitution for softwood; best opening face, which allows for improved sawmill wood-using efficiency; structural particleboard such as waferboard and oriented strand board; and the powered backup roll in plywood production. For pulp and paper the technologies were press drying, which allows for the increased use of hardwood fiber in paperboard, and various innovations permitting increased use of hardwoods for various types of papers.

Table 6-1 presents the estimated savings in softwood inputs by the year 2000. The savings estimates range from a 10 percent to a 25 percent reduction in softwood input requirements in the year 2000. These savings translate

Table 6-1. Estimates of the Savings in Softwood Use by Adopting Various Technologies

	Savings (percent)	
Technologies	1990	2000
Lumber technologies		
Improved design and engineering	10	20
Hardwood substitution	10	15
Improved yields from roundwood	4	10
Plywood technologies		
Hardwood substitution (waferboard, oriented strand board)	5	10
Improved yields from roundwood	5	10
Pulp and paper technologies		
Hardwood substitution (groundwood)	10	25
Hardwood substitution (printing paper)	10	15
Hardwood substitution (tissue and sanitary)	5	10
Hardwood substitution (bleached bristol)	10	15
Hardwood substitution (paperboard)	10	25

Source: Reprinted, with permission, from John Haygreen, Hans Gregersen, Irv Holland, and Robert Stone, "The Economic Impact of Timber Utilization Research," *Forest Products Journal* vol. 36, no. 2 (February 1986), pp. 12–20.

into average annual softwood savings rates of between 0.5 percent and 1.2 percent. Alternatively viewed, even if demand for the various final products were to increase by 10 to 25 percent over the period from 1980 to 2000, the demand for softwood input would remain unchanged over that period, as technological change would allow for both wood savings and hardwood substitution. Furthermore, the authors of the study (Haygreen and coauthors, 1986) maintain that they were very conservative in their estimates (pp. 15–16) and that the actual rates of saving could be significantly greater.

These findings reinforce the general premise that technological change can have and in fact is having a substantial effect on the long-term industrial wood supply. It is clear that in light of the current plentiful availability of the hardwood resource, much of the research and development effort will be directed at improving the ability to use the hardwood resource as a substitute for the higher-priced softwood resource. Hence the effort will be toward wood-extending as well as wood-saving technologies.

Although the innovations just examined form only a subset of all industrial wood processes, they do constitute a large and representative subset. Therefore, it would not be unreasonable to assume that an average annual rate of change in all wood technology would range between 0.5 and 1.2 percent. This would imply an increase in long-term wood availability at a given real price both through the wood-saving and wood-using effects of such change.

Wood-saving technologies enter the formal analysis of the TSM through the demand side, where, on the basis of recent historical trends as discussed in chapter 3, the growth in demand for industrial wood is posited to be modest—in large part because of the dampening influence of the wood-conserving effects of technological change.

THE EFFECTS OF TECHNOLOGICAL CHANGE ON INDUSTRIAL WOOD PRICES

Technological change, whether manifested in yield-enhancing more rapid growth of the timber resource or in the form of a wood-saving or wood-extending technology, can be expected to have the effect of placing downward pressure on the relative price of the wood resource. In an interesting follow-on to the study of Haygreen and his coauthors, Skog and Haynes (1987) incorporated the eight wood-saving and wood-extending innovations into the Forest Service's Timber Assessment Market Model to estimate their effect on the time profile of future harvests and market prices. Stressing that their findings were indicative rather than definitive, the authors found that the adjustment of the TAMM projections to account for the eight innovations caused softwood stumpage prices to be 30 percent lower in the year 2000 than in the absence of the innovations; this occurred while softwood timber harvests were simultaneously decreased by 4 percent. Hardwood, which was substituted for softwood in many of the innovations, had stumpage prices that increased 90 percent (however, the hardwood stumpage price was still below the softwood stumpage price). Yet its harvest level increased 7 percent above what it would have been in the absence of the innovations. More generally, the model projected that the price differential between softwood and hardwood stumpage would decline markedly as technology allowed the plentiful and cheap hardwood resource to become a better substitute for softwood in many uses.

These findings are consistent with the view that over the long term, technology will tend to allow production to adapt to a broader array of species and fiber qualities and that technology will tend to have a dampening effect on future price rises, as it allows production to shift toward the utilization of abundant sources—wood or nonwood—and away from scarce, high-priced resources.

SUMMARY

Historical experience makes it clear that substantial technical change has taken place in the overall economy, in specific sectors such as agriculture, and also in the forest resource and forest processing industries. This chapter argues not only that the role of technology is important in forestry but also that long-term projections often fail to incorporate properly the effects of technological change. Such forecasts are likely to overestimate the growth of demand and underestimate the growth of supply. Projections generated without proper consideration of changing technology are therefore systematically biased toward overestimating scarcity and projecting rises in real prices.

Technological change that can affect the forest industry may take several forms. First, innovation external to the industry can have major effects on the industry—if, say, the external innovation involves the development of a product that is competitive with the products of the industry. One example is the development of innovations that allowed for the expanded use of fossil fuel, which resulted in the decreased use of fuelwood. Similarly, the development of electronic information systems surely will have profound effects on the paper industry. Second, technological change in processing and new products in the forest products industry often have wood-saving or wood-extending results. Third, silvicultural and biological innovations are occurring that increase the growth rate and/or quality of the wood resource. Fourth, technological change can occur in harvesting and thereby increase the accessibility and economic supply of existing stocks of timber resources.

The effects of technological change can be incorporated into the TSM's supply and demand functions. New non-wood-using products that replace wood-using products will have the effect of retarding growth in demand and are captured in the TSM through the specification of demand. Similarly, wood-saving technological change might be incorporated into demand, whereas wood-extending, wood-growing, and harvest-enhancing technologies are best incorporated into the model via the supply function.

Estimates of technological change from various sources suggest annual rates of change for the overall economy of about 1.8 percent and for the agricultural sector of about 1.5 percent. Much less work has been done in the forest sector; however, recent work suggests that productivity changes in softwood timber are growing in the U.S. South at a rate of about 0.5 to 0.8 percent yearly. Studies of the wood-processing sector suggest annual productivity changes of about 1.75 or 1.90 percent. Hence studies of the wood sector tend to be roughly comparable with what is known about productivity growth elsewhere in the U.S. economy, with productivity growth in the timber resource lagging behind that of the economy as a whole and that of the wood-processing sector in particular. This is predictable, as research in tree growing is fairly new, and the large stock of older trees will reflect only very gradually the effects of technological change on silviculture.

The foregoing summary suggests that technological change has been, is currently, and almost certainly will continue occurring in a variety of products and processes related to the forest products sector and therefore will profoundly affect this sector. In this study the effect is incorporated formally into the TSM on both the demand and the supply side. On the demand side it is incorporated by retarding the rate over time of the outward shift of the demand curve (chapter 7). Because the historical growth of timber harvests should capture the historical effects of wood-saving technological change, the rate of growth of the demand curve was determined to a large degree by the historical time path of actual demand, which shows the gradual moderation of demand for the wood resource over the postwar period.

On the supply side technological change is reflected in more rapid shifts of the supply curve than would be expected on the basis of the size, quality, and current accessibility of the forest inventory observed. In addition to the introduction of new areas of plantation forests, two adaptations were made in the TSM to attempt to capture the effects of technological change. First, a factor for yield-enhancing change was built into the projections to a maximum of 0.5 percent annually, entering the system through the planting of seedlings epitomizing such technology. This growth factor is considerably more modest than that estimated by Newman (1986) (see appendix L). Second, in many cases the existing inventory was assumed to be economically usable even when the species was not the regionally preferred species. This approach assumes the ability of technology to utilize less desired species and stands.

REFERENCES

Capalbo, Susan M., and Trang T. Vo. 1988. "A Review of the Evidence on Agricultural Productivity and Aggregate Technology," in Susan M. Capalbo and John M. Antle, eds., *Agricultural Productivity: Measurement and Explanation* (Washington, D.C., Resources for the Future).

Cubbage, Frederick W., Bryce J. Stokes, and James E. Granskog. 1988. "Trends in Southern Forest Harvesting Equipment and Logging Costs," *Forest Products Journal* vol. 38, no. 2 (February) pp. 6–10.

Hayami, Yujiro, and Vernon W. Ruttan. 1971. *Agricultural Development: An International Perspective* (Baltimore, Md., The Johns Hopkins University Press).

Haygreen, John, Hans Gregersen, Irv Holland, and Robert Stone. 1986. "The Economic Impact of Timber Utilization Research." *Forest Products Journal* vol. 36, no. 2, pp. 12–20.

Manthy, Robert S. 1978. *Natural Resource Commodities—A Century of Statistics* (Baltimore, Md., The Johns Hopkins University Press for Resources for the Future).

Newman, David H. 1986. "Changes in Southern Softwood Productivity: A Modified Production Function Analysis," Working Paper no. 29 (Research Triangle Park, N.C., Southeastern Center for Forest Economics Research).

Olson, Sherry H. 1971. *The Depletion Myth: A History of Railroad Use of Timber* (Cambridge, Mass., Harvard University Press).

Papertree Economics Ltd., 1986. "Special Report" (January) (Vancouver, British Columbia, Papertree Economics Ltd.).

Potter, Neal, and Francis T. Christy, Jr. 1962. *Trends in Natural Resource Commodities* (Baltimore, Md., The Johns Hopkins University Press for Resources for the Future).

Risbrudt, Christopher D. 1979. "Past and Future Technological Change in the US Forest Products Industry," Ph.D. dissertation (East Lansing, Michigan State University).

Robinson, Vernon L. 1975. "An Estimate of Technological Progress in the Lumber and Wood Products Industry," *Forest Science* vol. 21, pp. 149–154.

Sedjo, Roger A. 1983. *The Comparative Economics of Plantation Forestry: A Global Assessment* (Washington, D.C., Resources for the Future).

Solow, Robert M. 1956. "A Contribution to the Theory of Economic Growth," *Quarterly Journal of Economics* vol. 70 (February) pp. 65–94.

———. 1957. "Technical Change and the Aggregate Production Function," *Review of Economics and Statistics* vol. 39, no. 3, pp. 312–320.

———. 1965. *Capital Theory and the Rate of Return* (Chicago, Rand McNally).

Skog, Kenneth E., and Richard Haynes. 1987. "The Stumpage Market of Timber Utilization Research," *Forest Products Journal* vol. 37, no. 6, pp. 54–60.

Styan, G. 1980. "Impact of North American Timber Supply on Innovations in Paper Technology," *Paper Trade Journal* vol. 164 (May).

U.S. Congress, Office of Technology Assessment (OTA). 1984. *Wood Use: U.S. Competitiveness and Technology*, vol. II, Technical Report OTA-M-244 (Washington, D.C., OTA).

Williams, Douglas, and Robert Gasson. 1986. "The Economic Stock of Timber in the Coastal Region of British Columbia" (Vancouver, University of British Columbia, Forest Economics and Policy Analysis Project).

7

A Technical Presentation of the Timber Supply Model

In chapter 4 we presented a broad nontechnical discussion of the Timber Supply Model that is used extensively in this study. This chapter provides a concise formal presentation of the model that together with appendix O should provide the technical reader with a detailed understanding of the TSM's technical features.

One focus of this chapter is price as the driving force behind the harvests and evolution of the forests. In addition, the role of the backdrop land classes is developed as well as that of investments in regeneration. Finally, the model is used to provide information about the supply function for timber. Because harvests are in essence an inventory adjustment, current harvests have an effect on future harvest levels. Hence current timber supply is related in part to expectations about future events, whereas future supply is constrained by current harvests.

THE MODEL: AN INTRODUCTION

In this study we discuss a significant portion of the world supply of timber; hence the forestry problem that we model abstracts from some issues in order to concentrate on specific issues and to keep the problem manageable. To focus on rather broad questions concerning timber supply without delving into issues of the composition of forest products, we treat such products as a single commodity. This and other simplifying assumptions aid us in modeling and discussing the broader issues; however, they also prevent us from examining directly some interesting questions. For example, as the price of timber rises, inferior grades are drawn into the market. They can enter either as substitutes in final demand or as processed substitutes. In addition, the rate of technological change for the processing of these substitutes is expected to be related directly to the value of timber. One way that we approximate this effect is to examine scenarios with different types and rates of technological change.

The portion of the world that we model is composed of twenty-two land classes in seven regions. Each land class has its own biological yield function that relates inputs to output. The output is merchantable volume, and the inputs are hectares of land of a specific quality experiencing certain average weather conditions, silvicultural or management practices called regeneration inputs, and aging of the trees. Each land class has its own yield function. In addition, the land base is exogenously determined, with each land class having a fixed land area except for the land class of the emerging region, which has an increasing land base over an initial time period. The effect of different rates of increase of the emerging region's land base are examined by exogenously varying that rate to generate different scenarios.

Certain harvesting and transportation costs vary by land class. The harvesting and local transportation costs are highest for those land classes with the roughest average terrain, and the international transportation costs are highest for the land classes that are farthest from an international market.

A DESCRIPTION OF THE MODEL

The model is an extension of our supply-potential optimal-control model, presented in *Forest Science* (Lyon and Sedjo, 1983). The objective function for the model is the discounted present value of the time stream of net surplus (consumers' plus producers') resulting from the harvests. This objective was selected because it maximizes the total benefit to society. The time path of harvests generated with it is economically efficient from society's viewpoint, and the solution time paths are the same as those generated by price-taker firms (Just and Hueth, 1979). That is, at the margin the resources used to produce the timber, including the land and time to age the trees, have a value in use equal to their opportunity cost to society. This function is maximized subject to the initial conditions and rules governing the evolution (laws of motion) of the system.

An alternative objective function commonly used in forestry models is the discounted present value of the net income stream of the landowners. This is the appropriate objective for individual landowners but not for society as a whole. The use of this objective function in our model would yield a worldwide monopoly or price-searcher's solution, which would not be optimal from society's viewpoint. (See Rahm [1981] for a similar discussion.)

Net surplus in year *j* is the area under year *j*'s demand curve from zero to the volume harvested minus harvesting, transporting, and regeneration costs for year j, where the costs are the sum of costs over the twenty-two land classes. Each land class has its own cost function, which is determined by specific features such as harvesting terrain, log size, accessing terrain, distance to the mill, and distance to the international market. The costs for a land class in year j will depend on the cost function, the volume harvested, and the hectares of land harvested and regenerated in the land class in year j.

The volume harvested for each land class depends on the yield function for the land class and the hectares by age group of trees harvested. The yield function for a land class is determined by characteristics such as climate, terrain, and soil quality; the yield per hectare is a function of the age of trees (years since the trees were regenerated) and the intensity of management practices associated with the hectare of trees.

The initial conditions are relevant items from history, including such items as hectares of forest by age group and land class as well as the level of the composite regeneration input for each of the hectares. This composite input is the present value of all planting and silvicultural operations, including precommercial thinning. We refer to this as the level of regeneration input or as the intensity of forest management.

The laws of motion for the system are the rules that govern the system, including rules for aging and regenerating the forest. Such laws redefine hectares of trees in age group i in time period j to be hectares of trees in age group $i + 1$ in time period $j + 1$, and they regenerate each harvested hectare in the time period during which it is harvested. To simplify the discussion we call age group i trees i years old. However, this may not be descriptively accurate because of regeneration lag, which depends on the land class involved and the level of regeneration input applied to these hectares. The laws of motion also redefine the level of regeneration input associated with age group i in time period j as the level associated with age group $i + 1$ in time period $j + 1$ in order to keep together each hectare of trees and their associated level of regeneration input.

The choice (control) variables are for each year the hectares harvested by age group and land class, and the level of regeneration input for each land class. Selection of these determines the rotation age by land class, volumes harvested, and costs in each year. In addition, with the laws of motion and the initial conditions these variables determine the structure of the forest, hectares of forest, and regeneration input by age group and land class in each year.

We structure the problem so that it evolves to the stationary state; hence the computer program of optimization first solves for the stationary-state solution values of the control and state variables. It solves for the optimal length of the rotation period, the stumpage (net) price of timber, and the volume of timber harvested by land class in the stationary state. Then the optimal time profiles of these same variables are calculated for the transition period. This is done by solving the difference equation problem identified by the laws of motion, the first-order conditions, the initial conditions, and the terminal conditions, where the terminal conditions are determined by the solution stationary state.

The role of discrete optimal control theory lies in identifying the laws of motion and the equations and equalities for the necessary conditions. These are used to identify the difference equation problem that is solved iteratively to solve the problem numerically.

FORMAL MODEL

The net surplus in year j can be written as

$$s_j = \int_0^{Q_j} D_j(n)dn - C_j$$

where Q_j is the quantity or volume of timber harvested in year j; $D_j(Q_j)$ is the inverse form of the demand function for industrial wood in year j; and C_j is the total cost (expenditures) in year j.

The total costs are the sum of harvest, access, domestic, and international transportation costs (CH_j) and regeneration costs (CR_j). Harvesting and transportation costs in year j depend on the total volumes harvested by land class, and regeneration expenditures depend on hectares harvested (regenerated) and level of input used.

We define x_{hj} to be a vector of hectares of trees in each age group for land class h in year j with elements x_{hij}. The subscripts h, i, and j are for land class, age group, and year, respectively; thus x_{hij} gives for land class h the number of hectares of age group i trees in year j. Let z_{hj} be the vector of state variables for the regeneration input, with z_{hij} the level of regeneration input associated with age group i in year j for land class h.

Next, we define u_{hj} to be the control vector of portions of hectares harvested. The elements u_{hij} denote for land class h the portion of the hectares of trees in age group i harvested in year j. Let w_{hj} be the level of regeneration input per hectare for those hectares regenerated in year j, and p_{wh} be the price of regeneration input for land class h.

The merchantable volume of timber per hectare for land class h in time period j for a stand regenerated i time periods ago depends on i and on the magnitude of the regeneration input used on this stand (z_{hij}). We denote this merchantable volume as follows:

$$q_{hij} = f_h(i, z_{hij}) \tag{7-1}$$

With these definitions the volume of commercial timber harvested from land class h in year j, Q_{hj}, is given by

$$Q_{hj} = u'_{hj} X_{hj} q_{hj} \tag{7-2}$$

where X_{hj} is a diagonal matrix using the elements of x_{hj}, and the total volume harvested in the responsive regions (those which we are modeling) is the sum of these over all land classes.

Harvest, access, and transportation costs for land class h are a function of the volume harvested in that land class:

$$CH_{hj} = c_h(Q_{hj}) \tag{7-3}$$

Regeneration cost for land class h in time period j is given by

$$CR_{hj} = (u'_{hj} x_{hj} + v_{hj})\, p_{wh} w_{hj} \tag{7-4}$$

where the inner product in parentheses gives the hectares harvested in land class h, v_{hj} is the exogenously determined number of hectares of new forest land in land class h (v_{hj} is nonzero only in the emerging region), and the product of the last two terms gives expenditure per hectare.

We write the objective function, the present value of the net surplus stream, as

$$S_0(x_0, z_0, u, w) = s_0 + \rho s_1 + \ldots + \rho^{J-1} s_{J-1} + \rho^J S_J^*(x_J, z_J) \tag{7-5}$$

where ρ is the discount factor, $\exp(-r)$, with r the market rate of interest; J is the last time period; u is any admissible set of control vectors u_0, u_1, . . . , u_{J-1} (including all land classes); w is any set of admissible control scalars w_0, w_1, . . . , w_{J-1} (also covering all land classes); and $S_J^*(\cdot)$ is the optimal terminal value function.

Equation (7-5) is maximized subject to the laws of motion and constraints on the values of the control variables. The portions of hectares harvested are constrained to be nonnegative and less than or equal to 1, and the regeneration inputs are constrained to be nonnegative:

$$0 \leq u_{hij} \leq 1 \quad \text{for all } h, i, j \tag{7-6a}$$

$$0 \leq w_{hj} \quad \text{for all } h, j \tag{7-6b}$$

The laws of motion for the system are given by

$$x_{h,j+1} = (A + BU_{hj})x_{hj} + v_{hj}\, e \quad \text{for all } h, j \tag{7-7a}$$

$$z_{h,j+1} = Az_{hj} + w_{hj}\, e \quad \text{for all } h, j \tag{7-7b}$$

where

$$A = \begin{bmatrix} 0 & 0 & 0 & 0 & . & . & . & 0 \\ 1 & 0 & 0 & 0 & . & . & . & 0 \\ 0 & 1 & 0 & 0 & . & . & . & 0 \\ 0 & 0 & 1 & 0 & . & . & . & 0 \\ 0 & 0 & 0 & 1 & . & . & . & 0 \\ . & . & . & . & . & . & . & . \\ . & . & . & . & . & . & . & . \\ 0 & 0 & 0 & 0 & 0 & 0 & 1 & 0 \end{bmatrix} \qquad e = \begin{bmatrix} 1 \\ 0 \\ 0 \\ 0 \\ . \\ . \\ . \\ 0 \end{bmatrix}$$

$$B = \begin{bmatrix} 1 & 1 & 1 & 1 & . & . & . & 1 \\ -1 & 0 & 0 & 0 & . & . & . & 0 \\ 0 & -1 & 0 & 0 & . & . & . & 0 \\ 0 & 0 & -1 & 0 & . & . & . & 0 \\ 0 & 0 & 0 & -1 & . & . & . & 0 \\ . & . & . & . & . & . & . & . \\ . & . & . & . & . & . & . & . \\ 0 & 0 & 0 & 0 & 0 & 0 & -1 & 0 \end{bmatrix}$$

A, B, and U are M-square matrices; U_{hj} is a diagonal matrix using the elements of u_{hj}; and e is an M-vector where M is equal to or greater than the index number of the oldest age group in the problem.

The product Ax_{hj} moves x_{hij} to $x_{h,i+1,j+1}$. Each year each age group becomes older by one year. The product $BU_{hj}x_{hj}$ subtracts from the redefined quantities the hectares harvested and places them in the one-year-old category (the newly regenerated category). The exogenously determined hectares of new forest plantation in time period j, $v_{hj}e$, add these hectares to the one-year-old age group in time period $j + 1$. These can be expressed as

$$x_{h1,j+1} = u'_{hj}x_{hj} + v_{hj} \qquad \text{for all } h, j \qquad (7\text{-}8a)$$

$$x_{h,i+1,j+1} = x_{hij} - u_{hij}x_{hij} \qquad (i = 1, 2, \ldots, M - 1) \qquad (7\text{-}8b)$$

In the law of motion for z, the product Az_j moves z_{hij} to $z_{h,i+1,j+1}$. This parallel redefining of x and z keeps the regenerated hectares and the level of their regeneration input in the same relative position in their respective state vectors. The scalar product $w_{hj}e$ places w_{hj} in location $z_{h1,j+1}$. Thus

$$z_{h1,j+1} = w_{hj} \qquad \text{for all } h, j \qquad (7\text{-}9a)$$

$$z_{h,i+1,j+1} = z_{hij} \qquad (i = 1, 2, \ldots, M - 1) \qquad (7\text{-}9b)$$

Necessary Conditions

The discrete-time, optimal-control literature contains extensive general discussions of the discrete-time maximum principle (Abadie, 1970; Butkovskii, 1963; Dyer and McReynolds, 1970; Jackson and Horn, 1965; Katz, 1962; and Polak, 1971). In this section we state the maximum principle for this forestry problem and examine the necessary conditions. In the next section we link this information to the shooting (binary search) method, which is used in the solution algorithm. This procedure identifies the margins on which the equations hold and also the specific difference equations that are used to solve the problem numerically.

The maximum principle is a theorem that states that the constrained maximization of equation (7-5) can be decomposed into a series of subproblems. In each time period the following Hamiltonian is maximized[1] with respect to u_{hj} and w_{hj} subject to constraints (equations 7-6a, 7-6b, 7-7a, 7-7b).

The Hamiltonian for year j is

$$H_j = \int_0^{Q_j} D_j(n)dn - C_j + \sum_h \lambda'_{h,j+1}[(A + BU_{hj})x_{hj} + v_{hj}e] + \sum_h \psi'_{h,j+1}(Az_{hj} + p_{wh}w_{hj}e) \quad (7\text{-}10)$$

where

$$\lambda_{hj} = \rho[dS_j^*(x_j, z_j)/dx_{hj}] \qquad (j = 1, \ldots, J) \quad (7\text{-}11a)$$
$$\lambda_{hj} = \rho[(ds_j^*/dx_{hj}) + (A + BU_{hj}^*)'\lambda_{h,j+1}] \qquad (j = 1, \ldots, J-1)$$

In addition,

$$\psi_{hj} = \rho[dS_{hj}^*(x_j, z_j)/dz_{hj}] \qquad (j = 1, \ldots, J) \quad (7\text{-}11b)$$
$$\psi_{hj} = \rho[(ds_j^*/dz_{hj}) + A'\psi_{j+1}] \qquad (j = 1, \ldots, J-1)$$

The derivatives with respect to vectors are gradient vectors, and $S_{j+1}^*(\;)$ is the solution function in $j + 1$. The solution function in year $j + 1$ can be conceptualized as the result of an application of Bellman's optimality principle and backward recursion. The λ_{hj} and ψ_{hj} are costate (adjoint) vectors

[1] In general the necessary conditions require only that the Hamiltonian be stationary (Jackson and Horn, 1965, p. 390); however, a stationary value will be a maximum subject to the constraints because the constraints are linear and the Hamiltonian is quasi-concave at a stationary point.

and identify the shadow values of the hectares of forest and the regeneration input, respectively, in each age group in year j.

The necessary conditions for the constrained maximization of the Hamiltonians in equation (7-10) are both necessary and sufficient for the constrained maximization of equation (7-5). The correspondence of necessary conditions is the essence of the maximum principle (Halkin, 1966). The conditions are sufficient because an equivalent form of the constrained maximization of equation (7-5) can be shown to be the maximization of a quasi-concave function subject to a set of linear constraints.

The Lagrangian function and the Kuhn-Tucker conditions of this optimization problem are

$$L_j^H = H_j + \sum_h \zeta'_{hj}(1 - u_{hj}) \tag{7-12}$$

$$dL_j^H/du_{hj} = [D(Q_j) - c'_h(Q_{hj})]X_{hj}q_{hj} - x_{hj}p_{wh}w_{hj} + X'_{hj}B'\lambda_{h,j+1} - \zeta_{hj} \leq 0 \quad \text{(for all } h) \tag{7-13a}$$

$$(\partial L_j^H/\partial u_{hij})u_{hij} = 0 \quad \text{(for all } h \text{ and } i) \tag{7-13b}$$

$$\partial L_j^H/\partial w_{hj} = -u'_{hj}x_{hj}p_{wh} + \psi_{h1,j+1} \leq 0 \quad \text{(for all } h) \tag{7-13c}$$

$$(\partial L_j^H/\partial w_{hj})w_{hj} = 0 \quad \text{(for all } h) \tag{7-13d}$$

$$dL_j^H/d\zeta_{hj} = (1 - u_{hj}) \geq 0 \quad \text{(for all } h) \tag{7-13e}$$

$$(\partial L_j^H/\partial \zeta_{hij})\zeta_{hij} = 0 \quad \text{(for all } h \text{ and } i) \tag{7-13f}$$

These Kuhn-Tucker conditions, the laws of motion for the state variables (equations 7-7a and 7-7b), and the laws of motion for the costate variables (equations 7-11a and 7-11b) identify a two-point boundary value problem that can be used to solve both theoretical and numerical problems.

The Analytic Solution

Equations (7-7a) and (7-7b) identify the method for calculating the values of state variables (hectares of forest by age group and stock of regeneration investment) in each year, given the values for the control variables in each year. These equations move the state variables forward over time.

Equations (7-11a) and (7-11b) identify the method for calculating the costate variables (shadow values of the state variables) in each year, given the values of the control variables in each year. This procedure calculates the costate variables starting at year J and moving backward through time to the present (year $j = 0$). Finally, equations (7-13a) through (7-13f) identify the method of finding the values of control variables in each year.

We manipulate equations (7-13a) and (7-13b) into a set of difference equations and use the difference equations in our solution algorithm and in our theoretical discussions of the solution time profiles. The elements of dL_j^H/du_{hj} in equation (7-13a) can be written as follows (see appendix O for details):

$$p_{hj} x_{hij} q_{hi} - x_{hij} p_{wh} w_{hj} + x_{hij}(\lambda_{h1,j+1} - \lambda_{h,i+1,j+1}) - \zeta_{hij} \leq 0 \quad \text{(for all } h \text{ and } i) \qquad (7\text{-}14)$$

where p_{hj} is the net price or stumpage price of timber for land class h. Stumpage price is equal to the market price of timber, $D(Q_j)$, minus the marginal harvesting, accessing, and transportation cost of timber, $c_h'(Q_{hj})$. This equation gives the marginal net surplus (net shadow value) of harvesting hectares of trees by age group and land class.

In equation (7-14) $\lambda_{h1,j+1}$ is the shadow value of trees that are one year old in year $j + 1$ (that is, trees regenerated in year j). Examination of equation (7-11a) indicates that its solution value is the discounted value of the actual harvest of these trees in the future. The costate variable $\lambda_{h,i+1,j+1}$ is the discounted value of age group hi from next year. For age group hk it can be written (see appendix O for details)

$$\lambda_{h,k+1,j+1} = \rho(\lambda_{h1,j+2} + p_{h,j+1} q_{h,k+1} - p_{wh} w_{h,j+1}) \qquad (7\text{-}15)$$

which states that the opportunity cost of harvesting k-year-old trees in year j is the discounted value of the trees that could be regenerated a year in the future plus the stumpage price of timber next year, multiplied by the merchantable volume of timber on that hectare one year in the future, minus the optimal regeneration expenditures on that hectare.

Combining this last equation and equation (7-14) yields

$$\begin{aligned} p_{h,j+1} = {} & [p_{hj} q_{hk} - p_{wh}(w_{hj} - \rho w_{h,j+1}) \\ & + \lambda_{h1,j+1} - \rho\lambda_{h1,j+2} - \zeta_{hkj}/x_{hkj}]/\rho q_{h,k+1} \end{aligned} \qquad (7\text{-}16)$$

The solution to this discrete-time, optimal-control model is the time paths of state variables, control variables, and costate variables, such that the laws of motion for state variables, equations (7-7a) and (7-7b); the laws of motion of costate variables, equations (7-11a) and (7-11b); the difference equation for the net price of timber (stumpage price) given in equation (7-16); and the remaining first-order condition, equations (7-13b) through (7-13f), are simultaneously satisfied. These form a two-point boundary-value difference equation problem.

CHARACTERISTICS OF THE SOLUTION

The economic characteristics of the solution time paths are of interest. As in the nonrenewable resource problem, the time path of price plays a very central role. Solow (1974, p. 3) calls the growth rate of net price "the fundamental principle of the economics of exhaustible resources." For exhaustible resources this growth rate provides the incentive to ration the resource over time, and in the forestry problem the growth rate of stumpage price and the growth rate of trees (merchantable volume of timber) provide this incentive. In the exhaustible resource problem with extraction costs independent of the resource stock, the growth rate of net price is equal to the interest rate and is constant until the resource is physically exhausted. In the forestry problem with an initial inventory of old growth forests, as in the problem we are examining, the growth rate of net (stumpage) price depends on the interest rate and the growth rate of the trees in the last age group harvested (marginal age group) in the current year. The growth rate of stumpage price initially may be greater than, equal to, or less than the interest rate (Lyon, 1981); however, over time as the old growth forests are harvested, the growth rate of merchantable volume of timber in the trees of the marginal age group will increase and the growth rate of stumpage price will decrease, until finally in the stationary state the stumpage price will be constant.

In generating optimal time paths the model adjusts on three margins: the rotation margin, the regeneration input margin, and the land margin. The rotation margin is central to the foregoing discussion of the growth rate of net price, because as the old growth forests are harvested the rotation age decreases. The regeneration margin results from adjusting the level of management practices to the point where the discounted value of the marginal product is equal to the discounted cost of a unit of the input. Finally, the land margin exists because at low market prices of timber, some land classes have a negative stumpage price, which implies zero harvests from these land classes at these prices; however, as the market price of timber rises, the stumpage price for one or some of these land classes may become zero or positive, drawing it or them into the harvest. We refer to these as backdrop land classes because they are the backdrop to prevent or moderate the price rise.

The solution time paths of harvests and market price contain some information for timber supply curves, and comparative static analysis can be used to generate additional information for these curves. A point (price and volume) on the supply curve for each year can be obtained from the solution time paths. To generate additional information about the supply curve for a particular year—say, the first year—the demand curve for that year can be shifted, with all other exogenous elements of the problem held constant. Those items held constant include the initial inventories, demand curves in all other years, exchange rates, yield functions, and cost functions. The new solution will identify an additional point on the supply function for that year.

Price as the Driving Force

The solution time paths maintain optimal inventories of trees and management practices by age group and land class, and they extract optimal amounts from the inventories of trees in each year. This is simply a reflection of the linking that exists within each year and between years for solution time paths. The inventories that exist at each point in time are rationed between the current and future harvests; a decision is made for each hectare of trees in each year as to whether to harvest the trees or to age them for another year. The option that adds the most to the present value of net surplus is selected, and in the comparison both currently existing trees and future rotations on the hectare are taken into account.

The following equations are useful in discussing these harvest decisions. The first is derived from equation (7-14). It holds when at least some hectares of trees are harvested from age group i in land class h during year j. The second is equation (7-15); it holds when the age group just identified is exhausted in either year j or $j + 1$:

$$p_{hj}q_{hi} - p_{wh}w_{hj} + \lambda_{h1,j+1} - \lambda_{h,i+1,j+1} \geq 0 \qquad (7\text{-}17)$$

$$\lambda_{h,i+1,j+1} = \rho[\lambda_{h1,j+2} + p_{h,j+1}q_{h,i+1} - p_{wh}w_{h,j+1}] \qquad (7\text{-}18)$$

where

p_{hj} = stumpage (net) price per cubic meter of timber harvested from land class h in time j (market price per cubic meter of timber in year j minus marginal harvest, access, and transportation costs per cubic meter for land class h)

q_{hi} = yield of merchantable volume in cubic meters per hectare for age group i in land class h

p_{wh} = price per unit of regeneration input in land class h; equals 1 in the base case

w_{hj} = regeneration input per hectare for those hectares regenerated in land class h in year j; composite input calculated from the discounted value of all management practices applied to this hectare between this and the next harvest

$w_{h,j+1}$ = same as above except for those hectares harvested and regenerated in year $j + 1$

$\lambda_{h1,j+1}$ = shadow value of a hectare of trees in land class h regenerated in time j; discounted value of future harvests from this hectare if they are regenerated in year j (they are one year old in year $j + 1$)

$\lambda_{h1,j+2}$ = same as above except the hectare of trees is regenerated in year $j + 1$

$\lambda_{h,i+1,j+1}$ = shadow value in year j of a hectare of age group i trees in land class h if their harvest is deferred until year $j + 1$

ρ = discount factor, $\exp(-r)$, with r the market rate of interest

Equation (7-17) gives the net shadow value of harvesting a specific hectare of trees in year j (contribution to the present value of net surplus stream for harvesting the hectare of trees in year j rather than in year $j + 1$), with the first three terms giving the value if the trees are harvested and the last term giving the opportunity cost of harvesting the hectare. The value if harvested in year j consists of the receipts from selling the stumpage ($p_{hj}q_{hi}$) plus the discounted value of regenerated future harvests ($\lambda_{h1,j+1}$) minus expenditures to generate the trees of the next harvest from this hectare ($p_{wh}w_{hj}$). The opportunity cost of harvesting the hectare in year j ($\lambda_{h,i+1,j+1}$) is the value of a forgone opportunity. It is identified in equation (7-18) as the discounted value of harvesting these trees next year when they are one year older. The description of the three terms of this value is the same as that given above except that these terms are deferred one year into the future.

All hectares for which equation (7-17) is positive and at least some of the hectares for which equation (7-17) is equal to zero (marginal age group) in year j are harvested in year j. It is useful to order the hectares harvested in a year, listing first those with the highest net shadow value and last those with a zero net shadow value. The hectares at the head of the list are harvested first, as they make the largest net contribution to net surplus; those at the end of the list are harvested last, as they would make the same net contribution if harvested in the current year or the next year. The age group of the hectares harvested last is called the marginal age group, and it plays a central role in the determination and description of the solution time paths.

The optimal harvests satisfy equation (7-17) and allocate or ration the inventory of timber between years. These principles are involved in the stationary state and in the transition from an initial inventory of trees to the stationary state.

The economics of the transition from an old growth inventory and the economics of exhaustible resources contain some similar elements. In the simplest exhaustible resource model, the only problem is the rationing of a known amount of the resource over time in an optimal way. The optimal rationing of extractions is achieved through increases in the value of the resource over time, which is the result of an increase in net price at a rate equal to that of the rate of interest. At the margin extractions in every year make the same contribution to the present value of net surplus, because the value of extractions is growing at the same rate as the rate of discount used in the present value calculations. Thus price is the driving force behind the rationing and extractions.

In the current forestry model, price is also a driving force behind the rationing of the harvests from the inventories. This is particularly obvious during the transition period when old growth forests are being harvested. Whenever it is optimal to draw down the inventory of timber, the change in stumpage price plays an important role in rationing the inventory between current and future harvests; if the old growth inventories are sufficiently large, the change in stumpage price has the dominant role. (See Lyon [1981] for a related analysis of the role of price.)

The relationships involved can be made more concrete using equation (7-19), which is derived from equations (7-17) and (7-18) (see Lyon and Sedjo [1986] for a similar discussion). The equation is for the marginal age group k (last age group harvested) in year j:

$$\begin{aligned} p_{h,j+1} = & [p_{hj} q_{hk} - p_{wh}(w_{hj} - \rho w_{h,j+1}) \\ & + \lambda_{h1,j+1} - \rho \lambda_{h1,j+2}] / \rho q_{h,k+1} \end{aligned} \tag{7-19}$$

The stumpage prices in years j and $j + 1$ for the marginal age group are linked by this equation. In addition, because stumpage price is market price minus the sum of marginal harvesting and transporting costs, the equation links market price in years j and $j + 1$. That is, the marginal age group determines the growth of market price as well as the growth of its own stumpage price. Furthermore, our solution algorithm uses an iterative procedure to solve a difference equation problem that includes this one. By determining the relationship between price in years j and $j + 1$, this equation together with the demand function determines the relationship between harvests in these two years. In addition, by requiring that the price evolve to the stationary state price while requiring that the inventory of hectares of trees by age group evolve from the initial inventory to the stationary state inventory, this equation firmly establishes the importance of price as a driving force.

To examine the implications of equation (7-19) more closely, we define

$$\begin{aligned} R_j &= [\lambda_{h1,j+1} - p_{wh} w_{hj}] / q_{h,k+1} \\ R_{j+1} &= [\lambda_{h1,j+2} - p_{wh} w_{h,j+1}] / q_{h,k+1} \end{aligned}$$

and

$$q_{h,k+1} = \exp(\Gamma)\, q_{hk}$$

where Γ is the growth rate of merchantable yield for age group k in land class h. The variable R_j is the discounted value of the stumpage in the future harvests from this land minus the regeneration cost of the next stumpage, all divided by yield per hectare of trees $k + 1$ years old. Its dimensions are dollars per cubic meter, the same as stumpage price, and it identifies on a per-cubic-meter basis the net present value of the future rotations initiated in year j. It is the land rental per cubic meter of volume. The variable R_{j+1} is the same concept, with the regeneration starting in year $j + 1$. Using these and the definition of ρ (= $\exp(-r)$), equation (7-19) can be written as

$$p_{h,j+1} = \exp(r - \Gamma)p_{hj} + [\exp(r)R_j - R_{j+1}] \qquad (7\text{-}19')$$

The difference $\exp(r)R_j - R_{j+1}$ captures the effect of future rotations on the basis of cubic meters of volume. If the hectare is harvested in year j rather than $j + 1$, then this difference is earned. It represents the return to not delaying the regeneration of trees. Alternatively, it represents the net opportunity cost of putting off the regeneration of trees for one year. This difference can be written as $\hat{r}R_j - \Delta R_j$, where $\hat{r}$ is the annual compounding interest rate corresponding to the continuous compounding rate r, and ΔR_j is $R_{j+1} - R_j$. The term $\hat{r}R_j$ is interest earned on the land rental by not deferring the harvest one year. The term ΔR_j is the change in the land rental per cubic meter of volume; when positive it is part of the return to forgoing current harvests.

The growth rate of net price is $(r - \Gamma)$ when $\exp(r)R_j - R_{j+1}$ equals zero, which will occur when future rotations are nonexistent or of zero value and when the growth rate of R_j is equal to the rate of interest. The exhaustible resource result discussed earlier is the special case of equation (7-19′) where future rotations are nonexistent and Γ equals zero. The growth rate of net price during the transition period may be greater than the rate of interest. This would result when Γ is small, as when the marginal age group is old growth timber and the net opportunity cost of deferring future rotations is positive. During the transition from harvests of old growth forests to the stationary state, the growth rate of net price will start high, maybe above the interest rate, then gradually decrease to zero in the stationary state. (For further discussion of these points, see Lyon [1981].)

Backdrop Land Classes

When the stumpage price for the oldest age group of a land class is negative, the harvest from this land class is zero. If the market price stays sufficiently low that it remains negative, the trees will become (if they are not already) old growth trees. Over time as demand and market price increase, stumpage price for this land class may become zero or positive. When this happens this land class will be drawn into the harvests, which will have a dampening effect on the rise of market price. If the volume of old growth timber in this land class is large enough, the effect of this situation may be a complete halting of the rise in market price either temporarily or for an extended length of time. Because of these characteristics we call land classes with negative stumpage price backdrop land classes.

The backdrop land classes will operate with natural regeneration and zero regeneration input because of the low or negative stumpage price. Growth of price, however, can change this, making the application of management practices to these hectares economically attractive.

The model adjusts on several margins. We refer to the drawing in of backdrop land classes as an adjustment on the land margin.

Regeneration Input

The application of management practices to land will have an effect on the yield of merchantable volume. This application is modeled as a composite input that is the present value of all planting and silvicultural operations, including precommercial thinning. These expenditures are per hectare and are optimally allocated between time periods; that is, if the discounted value of the regeneration expenditure is held constant, output per hectare is maximized in those time periods which are candidates for the optimal rotation period.

For the U.S. land classes we measure the input in dollars, and we can define a unit of input to be the quantity that \$1 will buy. For these land classes, therefore, we can refer to the level of this input in either physical quantities or dollar amounts. This same principle holds for the non-U.S. land classes for scenarios using the base case, intermediate exchange rates. For these the price of the regeneration input is \$1 ($p_{wh} = 1$).

The optimal level of the regeneration input in year j will satisfy the equation

$$\rho^k p_{j+k}(\partial q_k/\partial w_{hj}) = p_{wh}$$

The hectares are regenerated in year j and harvested at age k in year $j + k$. The equation is satisfied if the discounted value of the marginal product (VMP) of the input is equal to the cost of the input. This discounting of the VMP is from the actual time period when the regenerated forest is harvested to the period when it is regenerated.

These management practices give the model an intensive margin that it can adjust. In those scenarios where the market price increases significantly, we see intensive management practices used on the more productive and lower-cost land classes.

SUPPLY CURVE FOR TIMBER

Up to this point in this chapter we have been using the TSM to discuss the optimal time profile of harvests. We now turn our attention to the comparative static applications of the model that yield information about the supply function for timber. This supply function is of interest not only because people frequently discuss and attempt to estimate it but also because the concepts involved are useful in understanding the economics of forestry.

The supply function for timber can be defined similarly to the conventional supply function; however, because timber harvests are from the inventory of standing timber, there are some fundamental conceptual differences. For the conventional supply function the long-run elasticity of supply depends primarily on the elasticities of supply for the inputs. The inputs or resources are competed away from the production of other commodities in the economy. In forestry, however, only the resources to harvest, transport, and regenerate the trees are competed away from other commodities. The primary input, the standing trees, must be competed away from harvests in other years.

The supply function identifies the quantity of a product that would be supplied per unit time or for a given time period at different prices under given conditions. With price-taker firms price is equal to the marginal opportunity cost of producing a unit of the product; hence it gives the relation between quantities and opportunity cost. The list of given conditions for the timber supply function for the current year includes the following:

1. The inventory of hectares of trees by age group, stocking rate, location, and land class.
2. The yield functions for commercial volume of timber as a function of age and management practices used for each land class.
3. Expected future levels of demand for timber.
4. Supply functions for purchased inputs used to harvest, transport, and cultivate the trees (supply functions for these activities).
5. The market rate of interest.
6. Expected future plantings of plantations.
7. Expected future technological change.

This list includes past events, production relationships, supply of inputs, and future events. The list for the conventional supply function would include all but future events.

A manipulation of equation (7-17) for the marginal age group is useful in identifying the concepts involved. Replacing stumpage price in equation (7-17) with market price, $D(Q_j)$, minus marginal harvest, access, and transportation costs, MC_h, yields the following:

$$D(Q_j) = MC_h + [p_{wh}w_{hj} - \lambda_{h1,j+1} + \lambda_{h,k+1,j+1}]/q_{hk} \qquad (7\text{-}20)$$

The market price per cubic meter of timber is equal to the marginal opportunity cost (MOC) of timber in the marginal age group (the equation's right side). This opportunity cost consists of the cost of the resources used to harvest and transport the last unit harvested and the forgone future value of the trees per cubic meter of timber harvested less the net future value of the regenerated trees. The terms can be explained as follows:

$\lambda_{h,k+1,j+1}/q_{hk}$ = shadow value of the timber per cubic meter if harvested in the future, or future value forgone by harvesting the trees in year j

$[\lambda_{h1,j+1} - p_{wh}w_{hj}]/q_{hk}$ = net shadow value of the trees that could be regenerated in year j per cubic meter of timber harvested, or present value of the trees that could be regenerated in year j minus their regeneration costs, or land rental per unit of volume

These two terms identify the effect of the future and are absent in conventional supply analyses.

The supply function for some year—say, year one—is the result of a comparative static analysis in the sense that other things are held constant and the volume harvested in year one is changed (say, increased) by some means. The foregoing list presents the items held constant. In general we would expect the MOC to be an increasing function of the volume harvested in year one; however, special cases can occur where the MOC is constant over at least some range. The marginal cost MC_h could be expected to be a nondecreasing function of volume harvested, and in our applications it is a constant function. Increases in current harvests cause decreases in future harvests because a smaller inventory of trees is transferred to the future. (A possible exception that yields no change in future harvests is discussed below.) The decrease in future harvests will cause shifts along the given future demand functions to higher prices, thereby causing an increase in the shadow value of timber in future harvests ($\lambda_{h,k+1,j+1}$). In addition, this decrease in future harvests will increase the shadow value of the regenerated trees ($\lambda_{h1,j+1}$); however, this increased value must be discounted over one rotation, and the effect that far in the future will be small compared to the effect over the next few years. Increased harvests in the current year may exhaust the current marginal age group, causing selection of the age group with the next lowest MOC. This would yield an increase in the MOC.

There may also be a minor effect on the intensive margin. The higher future prices will increase the optimum level of the regeneration input, which will increase the MOC. The modifier "minor" is used here because the effect comes through the shadow value of the regenerated trees that are harvested one rotation in the future. The chain of effects would be from increased current harvests to decreased harvests next year, which in turn would cause decreased harvests the following year, and so forth. The chained effect should be negligible at some distant future year.

The change in the optimal level of the regeneration input will be greater for the land classes with the shorter rotation periods; however, unless this land class contains the marginal age group, the foregoing discussion will not capture this effect. The more intensive management practices on land classes other than the one containing the marginal age group will increase future inventories, thereby moderating increases in future market prices and in the shadow value of timber in future harvests.

An exception to the smaller future harvests and increasing MOC can occur when there is a backdrop land class that is never exhausted in the sense of moving over time to an optimal rotation period. Additional current harvests will not have to compete with future harvests; therefore, the shadow value of the timber in future harvests for the marginal age group will be zero ($\lambda_{h,k+1,j+1} = 0$). The regenerated trees in the backdrop land class will similarly have a zero stumpage price. The return to management practices on this land class will be zero; therefore, management practices on the backdrop land class will also be zero. This implies that the MOC will equal MC_h; hence MOC will increase with the harvests only if MC_h does.

Another possibility is that the backdrop land class will have an effect over a limited range of harvests and then will be exhausted in the sense just discussed. The MOC would then rise with the possibility that another land class would become the backdrop. In this fashion the backdrop land classes can have a large effect on the elasticity of the supply function.

In summary, timber harvests are inventory adjustments. Changes in the inventory left for the future affect future harvests, thereby affecting the shadow value of trees in these harvests. The optimal adjustments that take place are at the rotation margin, the land margin, and the intensive margin. Adjustments at the rotation margin occur because the youngest age groups that are harvested change. Adjustments at the land margin occur as additional land classes are drawn in at higher market prices, and adjustments are made at the intensive margin as future prices rise. The more extensive the adjustments at the rotation margin as the current harvest is increased, the larger the effect on the MOC and the more inelastic the supply function. In contrast, the more extensive the adjustments at the other two margins, the smaller the effect on the MOC and the more elastic the supply function.

REFERENCES

Abadie, J. 1970. "Application of the GRG Algorithm to Optimal Control Problems," in J. Abadie, ed., *Integer and Nonlinear Programming* (Amsterdam, North Holland).

Burkovskii, A. G. 1963. "The Necessary and Sufficient Conditions for Optimality of Discrete Control Systems," *Automatic and Remote Control* vol. 24, no. 8, pp. 963–970.

Dyer, P., and D. R. McReynolds. 1970. *The Computation and Theory of Optimal Control* (New York, Academic Press).

Halkin, H. 1966. "A Maximum Principle of the Pontryagin Type for Systems Described by Nonlinear Difference Equations," *Journal of SIAM Control* vol. 4, no. 1, pp. 90–111.

Jackson, R., and F. Horn. 1965. "Discrete Analogues of Pontryagin's Maximum Principle," *International Journal of Control* vol. 1, no. 4, pp. 389–395.

Just, Richard E., and Darrell L. Hueth. 1979. "Welfare Measures in a Multimarket Framework," *American Economic Review* vol. 69, no. 5, pp. 947–954.

Katz, S. 1962. "A Discrete Version of Pontryagin's Maximum Principle," *Journal of Electronics and Control* vol. 12, no. 2, pp. 179–184.

Lyon, Kenneth S. 1981. "Mining of the Forest and the Time Path of the Price of Timber," *Journal of Environmental Economics and Management* vol. 8, no. 4, pp. 330–344.

Lyon, Kenneth S., and Roger A. Sedjo. 1983. "An Optimal Control Theory Model to Estimate the Regional Long-Term Supply of Timber," *Forest Science* vol. 29, no. 4, pp. 798–812.

Lyon, Kenneth S., and Roger A. Sedjo. 1986. "Binary-Search SPOC: An Optimal Control Theory Version of ECHO," *Forest Science* vol. 32, no. 3, pp. 576–584.

Polak, E. 1971. *Computational Methods in Optimization* (New York, Academic Press).

Rahm, C. M. 1981. "Timber Supply Analysis and Baseline Simulations," Module III-A, report to the Pacific Northwest Regional Commission (Vancouver, Wash.).

Solow, Robert M. 1974. "The Economics of Resources or the Resources of Economics," *American Economic Review* vol. 64, no. 2, pp. 1–14.

8

Base-Case Forecast and Analysis

This chapter presents the projections of the base case of the Timber Supply Model and uses an economic framework to analyze the projections within the context of the real world. Because the base case incorporates what are viewed as the most probable set of assumptions, we have judged it to be the most likely scenario, hence our most probable forecast. In addition to the aggregate projections of intertemporal price and harvest levels over a fifty-year period, the projections provide estimates of the changing regional structure of production in response to economic and biological forces. The base-case forecast then is modified by the introduction of different assumptions about demand growth. Both high-growth and low-growth demand scenarios are examined. The scenario projections are compared and contrasted with the forecasts of the base case. An analysis of the changes in projections of the model under the different assumptions provides insights into the behavior and responsiveness of the global timber supply system over the long term.

The assumptions used in the TSM for the base-case forecast are as follows:

1. World demand for industrial wood initially increases at a rate of 1 percent annually, gradually falling in successive years to zero after fifty years (see chapter 3).
2. The production of the nonresponsive region increases at a rate of 0.5 percent annually, gradually falling to zero at fifty years (see chapter 3).
3. Biotechnological change is assumed to shift the growth functions upward at a maximum rate of 0.5 percent annually, declining linearly in successive years to zero after fifty years. This rate is adjusted downward in proportion to the investment in regeneration (see appendix L).

4. New forest plantations are established in the emerging region at a level of 200,000 hectares per year for a thirty-year period (see appendix B).

5. The dollar exchange rate is assumed to remain at an intermediate level (as judged by recent experience) throughout the period of the analysis (see appendix J).

These assumptions are incorporated into the TSM and allow us to generate our base-case forecast. The focus of the analysis is on the seven major timber-producing regions of the world discussed earlier, which collectively produce about one-half of the total global supply of industrial wood. Although other studies have assumed a more rapid rate of growth of demand for industrial wood, the modest rate of growth of demand for industrial wood assumed here allows for a much more rapid growth in demand for final wood products. The differences in growth, as discussed in chapters 3 and 6, are accounted for by the introduction of wood-saving technological change.

The assumptions of (1) a gradual reduction of growth in the harvests of the nonresponsive regions to zero in year fifty, (2) a gradual decline in the effect of changing biotechnology on growth functions, and (3) a cessation of the establishment of new forest plantations after thirty years are technical conveniences designed to coordinate the supply-side assumptions of the model with those of the demand side when examining a fifty-year time period. The initial 0.5 percent annual rate of growth of the harvest levels of the nonresponsive regions appears to be reasonable and roughly consistent with recent experience. (The effects of modifications in this rate are examined later in the scenario analysis.) Similarly, the conservative initial rate of biotechnological change in the yield functions, 0.5 percent annually, is consistent with such changes in agriculture and seems reasonable. Finally, the base-case assumption that new industrial forest plantations will be established at a rate of 200,000 hectares per year in the emerging region appears sensible in light of the experience of the 1970s and early 1980s.

The assumption concerning the rate of plantation establishment may be challenged and is surely somewhat arbitrary and simplistic. An economic perspective would suggest that the rate of plantation establishment is determined at least in part by the expected level of future timber prices. Thus, for example, one might expect to see the high-plantation-establishment scenario associated with the high-demand scenario, as the latter will result in higher prices (other things being equal) and thus "induce" higher levels of investment in plantations. Nevertheless, the assumed rate of new plantation establishment for the base case appears to be consistent and reasonable in light of recent experience, as discussed elsewhere in the text (see chapters 1, 2, 4, and 9). Finally, although the assumption of biotechnological change could be greater or less, it does appear reasonable and in any event does not have a large effect on the projections over the time period examined.

Recognizing that some assumption concerning long-term exchange rates is required, we believe that the conservative "intermediate" exchange rate assumption used in this study is sensible. This assumption is also examined later through the use of scenario analysis.

THE FOREST IN TRANSITION AND CHANGING HARVESTS

The seven responsive regions are quite different in terms of existing initial conditions. For example, the Pacific Northwest, western Canada, and much of eastern Canada are still very largely old growth regions so that initially the economic question is one of determining the optimum rate of drawdown of that old growth timber within the context of a world production and trading system. The Nordic region has experienced regular cutting and forest management. The U.S. South has largely second-growth forest, much of which has come under active forest management only fairly recently. The emerging region has no old growth inventory and constitutes only an agricultural-type timber-growing regime with a multiple-year harvesting cycle. This study has assumed an average "representative" harvest rotation, but the actual rotations vary from seven to almost thirty years. Finally, the Asia-Pacific region, which has an old growth forest of tropical hardwood that is harvested by a selective logging system, requires an adaptation of the TSM to capture the essence of the thirty-five-year polycyclical selective logging harvest operations. In summary, although much of the total timber-producing region consists largely of old growth forests, other parts are essentially multiple-year agricultural regimes, and still others are somewhere between these extremes.

In this study we use the TSM to examine the optimal rate of drawdown of an old growth forest while simultaneously examining the drawdown and regrowth of secondary and plantation forests within a unified integrated system. Within this context the model generates a structure of regional harvest projections that reflect the relative comparative advantages in timber harvesting and timber growing in the various regions. Furthermore, because the projections are intertemporal and depend on the age distribution and volume of the forest inventory, the projections reflect relative changes in regional comparative advantage that occur over time as the stock of old growth declines and as the economics of timber growing gradually replaces the economics of harvesting as the dominant economic factor determining production.

As noted earlier, the seven major timber-producing regions are further subdivided into a total of twenty-two land classes, and data specific to each site class have been incorporated into the model. At relatively low prices some of the twenty-two land classes will not be economical to harvest. At higher prices these land classes are drawn into economic production. In addition, changes in harvests then depend on changes in the rate of inventory drawdown, changes in rotation length, and changes in intensity of management, as well as on the introduction of new land areas into industrial forest either through the creation of new plantations in the emerging region or through the drawing of inaccessible regions into the harvesting process. Finally, biotechnological change that allows for more rapid tree growth also affects long-term supply.

The Base-Case Forecast: Some Results

The results of the base-case forecast are shown in figures 8-1 through 8-4. Figures 8-1 and 8-2 present the long-term forecast of the time profile of the total harvest volume of the global forest system and the harvest time profile for each individual region. The projection of total harvest shows modest growth in the harvest level for the entire fifty-year period, with the total harvest level stabilizing toward the end of the period. The entire fifty-year period experiences a total harvest increase of about 30 percent.

During the period much of the remaining accessible old growth is drawn down, primarily in the Canadian forests, while maturing inventories are harvested in the Nordic region. The Canadian forests show an overall decline in production over the period, with western Canada declining substantially over the first thirty years but also making a significant comeback in harvests toward the end of the period. The Nordic region shows increased harvests as postwar regeneration begins to mature, but it falls back toward the end of the fifty-year period. The Pacific Northwest and the Asia-Pacific region maintain approximately their initial harvest levels throughout the fifty-year period. The U.S. South, which is the largest single producing region initially, gradually increases harvests even further as the managed and plantation stands reach maturity and the effects of more intensive management are manifested in higher yields.

Throughout the fifty-year period the emerging region becomes an increasingly important wood source as new forest plantations are established and subsequently harvested. By the end of the period this region is clearly second only to the U.S. South as a timber producer. Globally, most of the system-wide increase in harvest over the fifty-year period results from increased harvests in the U.S. South and the emerging region.

The Transition to the Foresters' Regulated Forest

Interestingly, although the overall harvest exhibits little volatility either in the transition or the steady state (no attempt is made in the projections to account for the inevitable short-term fluctuations associated with the business cycle), individual regions—both transitional and steady state—do experience

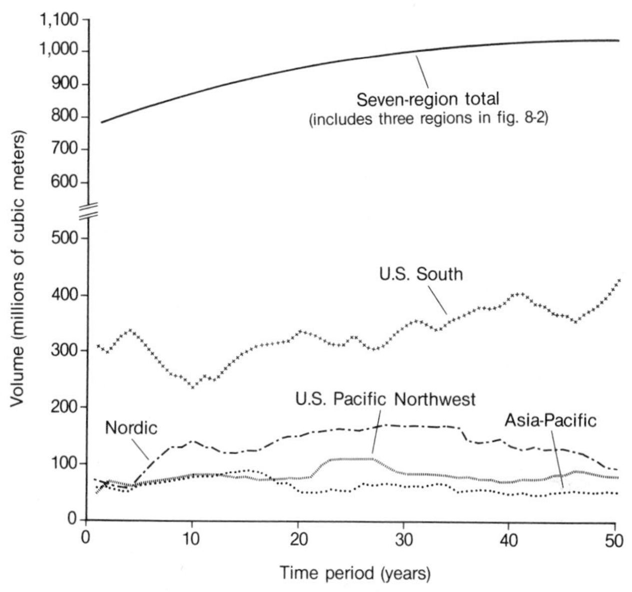

Figure 8-1. Harvest volume over time: base-case scenario (four regions).

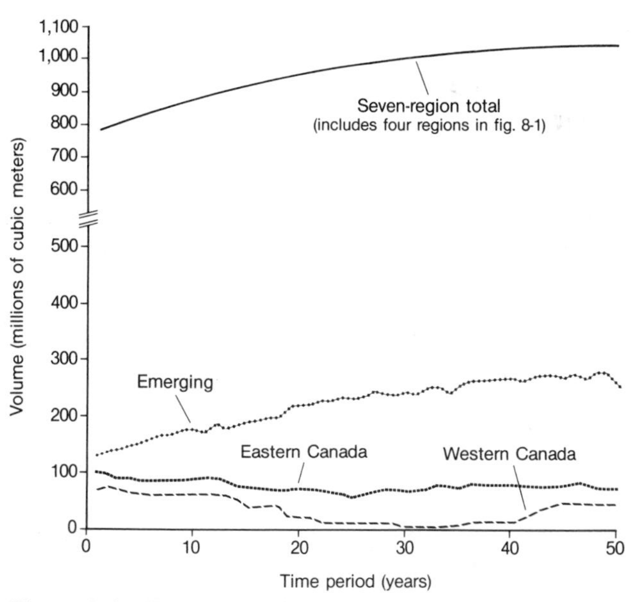

Figure 8-2. Harvest volume over time: base-case scenario (three regions).

substantial volatility throughout this period. This volatility reflects to a large degree the initial uneven age distribution of the forest; it tends to be greatest where rotations are long and the region begins the initial period with large volumes of old growth, and least where rotations are short and initial inventory is minimal. Hence the volatility is generally greatest in the traditional timber-growing regions of the Northern Hemisphere. The volatility is small in the Asia-Pacific region because of the polycyclical selective logging regime practiced, which in effect shortens the harvest rotation but not the growing rotation.

The steady state of the TSM differs in several important respects from the forester's notion of an "even-aged regulated stand." Rather than each individual forest approaching the foresters' ideal, the system approaches the ideal while leaving a considerable degree of volatility in each regional forest. This is not surprising when considered within the context of regional differences in species, different rotation lengths, and very different initial conditions. Hence, although the entire global forest system approaches the foresters' ideal of a steady-state regulated forest, individual forests behave as subparts of the total system with regional harvests that commonly wax and wane over decades.

The Changing Regional Structure of Harvests

An examination of each of the seven regions reveals that the U.S. South is forecast to be the major producer throughout the entire period; its harvest is somewhat variable but clearly exhibits a rising trend over the fifty-year period after an initial harvest decline. The emerging region gradually increases its production over the fifty years to become the system's second-largest producer after the U.S. South. The Nordic region and Canada are both projected to continue to be significant producers of industrial wood, with the Nordic region showing a high degree of volatility over time that largely reflects its forests' age distribution. Declines in Canadian harvests are substantial over the twenty- to forty-year period, particularly in the west, as the existing stock is drawn down and long growing periods are required before expansion occurs again as the new forests mature. The Pacific Northwest exhibits a fairly high harvest volatility over the long term. Furthermore, the Asia-Pacific region continues as a modest (by global standards) producer, showing somewhat less harvest volatility than most other regions.

An important finding of these projections is that the Northern Hemisphere old growth regions, such as western Canada and the Pacific Northwest, continue to be important suppliers of industrial wood throughout this fifty-year period and probably into the indefinite future, despite the growing role of the emerging region and the U.S. South. This phenomenon is in conflict with the often-expressed view that the temperate-climate timber-producing

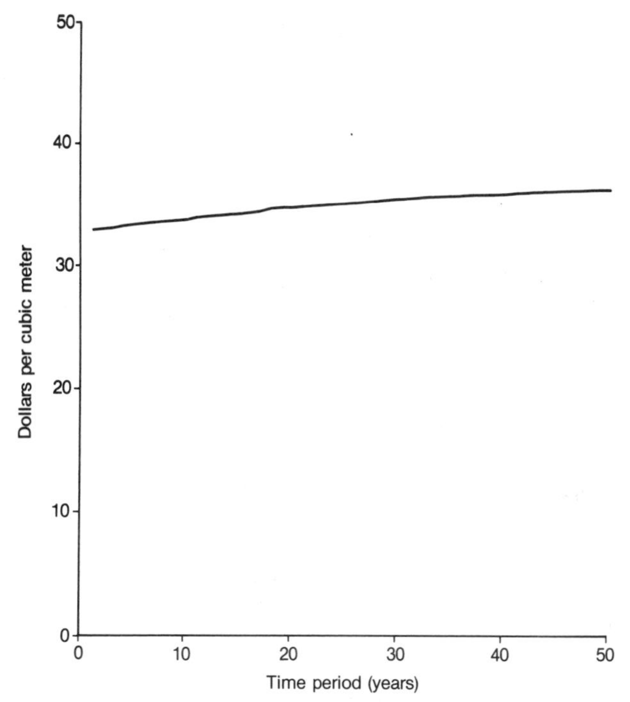

Figure 8-3. World market real price of industrial wood: base-case scenario.

regions cannot compete with the biologically superior timber-producing regions of the tropics and Southern Hemisphere. However, it should be noted that although no regions discontinue harvests in response to the competition provided by the increasing harvests of the plantation regions, there is a marked structural shift toward increased production harvests in the high-growth regions and a decline, sometimes relative but often absolute, in the role of the old growth temperate-region forests. As noted, the U.S. South and the emerging region account for all the long-term harvest increases and also offset harvest declines in some regions. This study suggests that of the seven regions examined in detail, British Columbia is perhaps the most vulnerable to harvesting declines in the face of increased competition from harvests elsewhere. However, other temperate regions are also affected.

Price Increases

Figure 8-3 presents the projection of the time profile of the world market real price. The projection of price shows only a very gradual rise for the base case, with the real price rising about 10 percent over a fifty-year

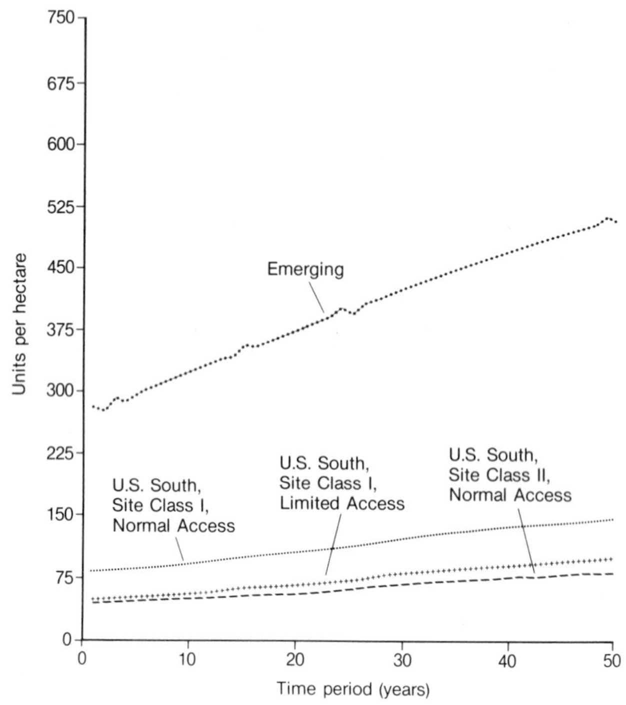

Figure 8-4. Regeneration investment: base-case scenario.
Note: Investment in the following regions was below 30 units per hectare: East Coast, Lake Forest, Normal Access; Pacific Northwest, West of Cascades, Normal Access; U.S. South, Site Class II, Limited Access; U.S. South, Site Class III, Normal Access.

period—or at an average annual rate of just under a very modest 0.2 percent. This forecast conflicts sharply with the conventional view in forestry that real prices will continue to increase rather strongly into the indefinite future.

Forest Investments and Management

Figure 8-4 presents projections of the time profiles of the optimal level of investment in regeneration intertemporally by land class for the base case. The modest intertemporal real price rises noted earlier obviously serve to moderate the level of investment projected in regeneration. In only eight of the twenty-two land classes examined are significant levels of investment in regeneration justified. The projections indicate that the emerging region justifies by far the highest levels of investment in regeneration per unit land area. Other regions where significant levels of investment in regeneration are justified are limited to five of the U.S. South's eight land classes, the good access sites on the west side of the Cascades in the Pacific Northwest region, and the Great Lakes region of eastern Canada. Regions and land classes not presented in figure 8-4 are projected to have only negligible levels of economically justifiable regeneration investment. For these regions economic forestry requires a high degree of reliance on renewal through natural regeneration; the projected low stocking levels and reduced future harvest reflect this situation.

Approximately 50 percent of the current systemwide total harvest comes from site classes that justify little investment in future regeneration. This harvest projection suggests that although intensively managed forests will provide an increased portion of the world's future wood supply, a very important future role remains for those forests which are not economic to manage intensively but rather rely largely on natural processes and low levels of management.

For land classes where artificial regeneration is justified economically, the optimum level of regeneration investment tends to increase over time, reflecting the fact that more distant future prices are somewhat higher than near-term prices. The concentration of investments on high-site lands demonstrates that the model's projections are consistent with the conventional wisdom in forestry, which suggests a positive relationship between site productivity and the level of forest management and regeneration investment.

It is worth stressing that these results obtain for the base case. As is demonstrated in the high-demand scenario analysis later in this chapter, a scenario that generates higher prices will also generate higher projected levels of investment in regeneration.

Unutilized Forest

Finally, although not revealed in the figures in this chapter, the base-case forecast results also reveal that three of the twenty-two potentially productive land classes do not harvest significant volumes of timber in the base-case forecast. This is because the real price does not rise to sufficiently high levels to justify the higher costs associated with harvesting in these less accessible and higher-cost timberland sites. In economic jargon these forests are submarginal at the projected prices. However, these unutilized forests constitute a reserve inventory of timber that could be utilized if the price rise becomes larger, as is demonstrated in the high-demand scenario developed later. In the real world, land classes that fit this example extend beyond the three formally incorporated in the model and could include, for example, the forests of Alaska, the nonoperative regions of Canada not included in our current formulation, the Amazon Basin, and the like.

CHANGES IN THE GROWTH OF FUTURE DEMAND

The base-case forecast, as noted, assumes a rather modest increase in demand for industrial wood over the next fifty years. We now examine the implications of both a substantially higher and a substantially lower rate of growth of demand for harvests, real prices, and investment in regeneration. This exercise affords an ideal opportunity to examine the manner in which the supply side of the TSM responds to changed stresses generated by different rates of demand growth. In principle an economic system has two types of response to changing demand—price and/or quantity adjustments. In the TSM both adjust.

The Demand Scenarios

Two alternative intertemporal rates of growth of demand are assumed. The high-demand scenario begins at a 2.0 percent growth rate, which is very close to the historical experience of 2.03 percent annually for the three-decade period from 1950 to 1980. However, as noted in chapter 3, this is viewed as a high-growth assumption for the future because the worldwide growth rate of demand declined substantially over the 1950–1983 period.

The low-demand scenario posits demand growing initially at 0.5 percent annually. This is below the actual rate of 1.1 percent experienced over the 1970–1980 decade. Both scenarios posit a demand growth that gradually declines in successive years to zero after fifty years. The decline is consistent with the post-1950 trend of globally declining growth of industrial wood consumption.

The High-Demand Scenario

The high-demand scenario posits the same conditions and assumptions as the base-case scenario except that world demand grows initially at 2.0 percent annually, declining gradually to zero at fifty years. It should be noted that symmetrical adjustments are not made on the supply side. Figures 8-5 and 8-6 present the projected time profile of the high-demand growth scenario's aggregate harvest levels for all the responsive regions and the individual harvest levels of each of the seven responsive regions. Under the high-demand scenario, the aggregate harvest of the responsive regions increases about 60 percent in the fifty-year period, or about 0.9 percent annually. In view of the more rapidly increasing demand, the real price experiences a sharp upward rise (figure 8-7), increasing 65 percent in the first fifty years—an annual growth rate of about 1.2 percent.

Not surprisingly, each of the seven responsive regions produces more harvests under the high-demand conditions than under the base case. The

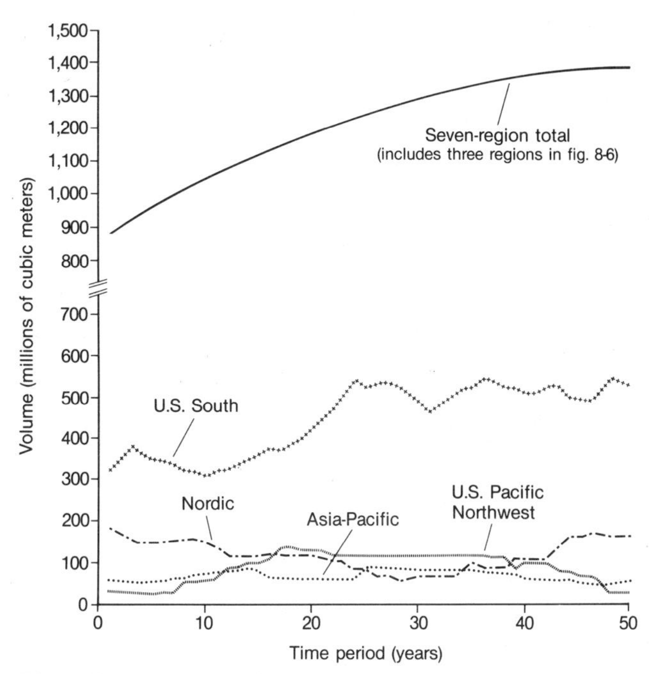

Figure 8-5. Harvest volume over time: high-demand scenario (four regions).

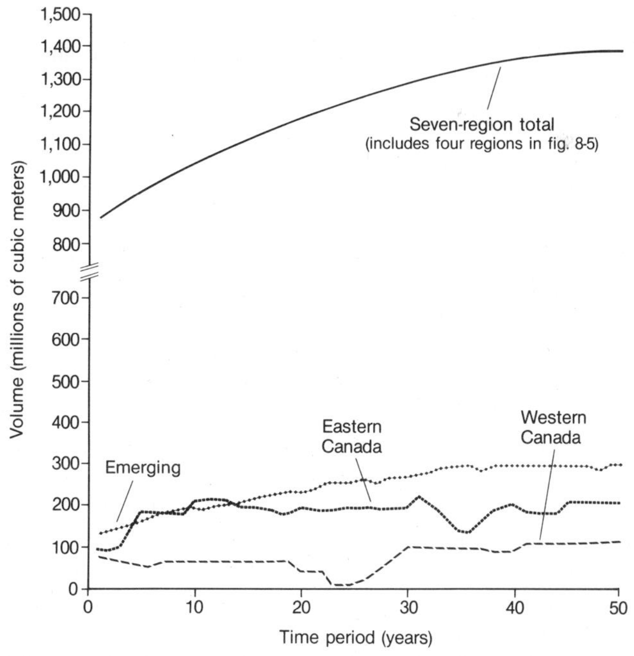

Figure 8-6. Harvest volume over time: high-demand scenario (three regions).

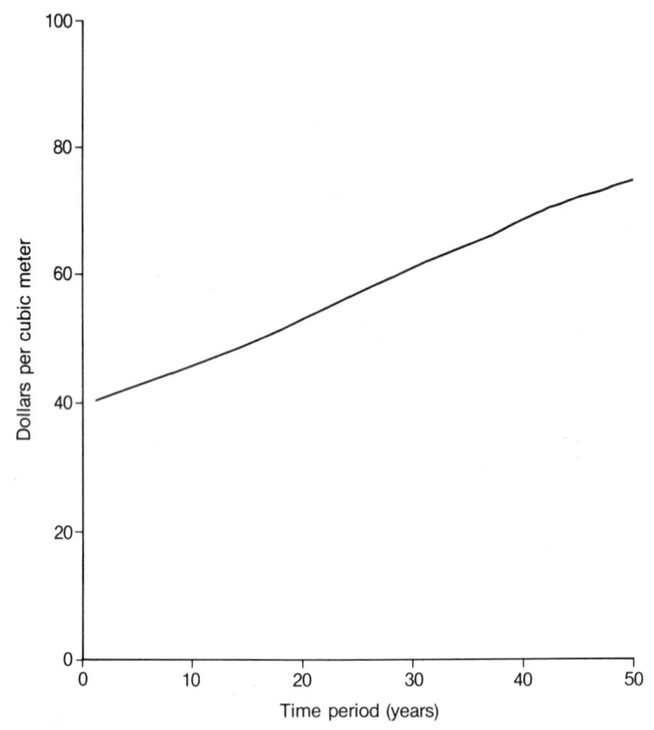

Figure 8-7. World market real price of industrial wood: high-demand scenario.

U.S. South increases its harvest by roughly 80 percent over the first forty years as it responds to the higher growth in wood prices with increased regeneration investments and wood from a previously excluded land class. Production in other regions also increases substantially. For example, Canadian harvests for both eastern and western Canada are much greater under the high-demand scenario than in the base case for both eastern and western Canada, largely because of the introduction of harvests from previously inaccessible land classes. Pacific Northwest harvests also increase. Harvests for these regions increase in part because the high prices now make logging on land classes with limited accessibility economically feasible, and hence draw into the harvesting solution all twenty-two land classes. In economic terminology, at higher wood prices the formerly submarginal land classes become supramarginal.

The introduction of the previously submarginal land classes also affects the time profile of harvests for certain regions. For example, harvests in the

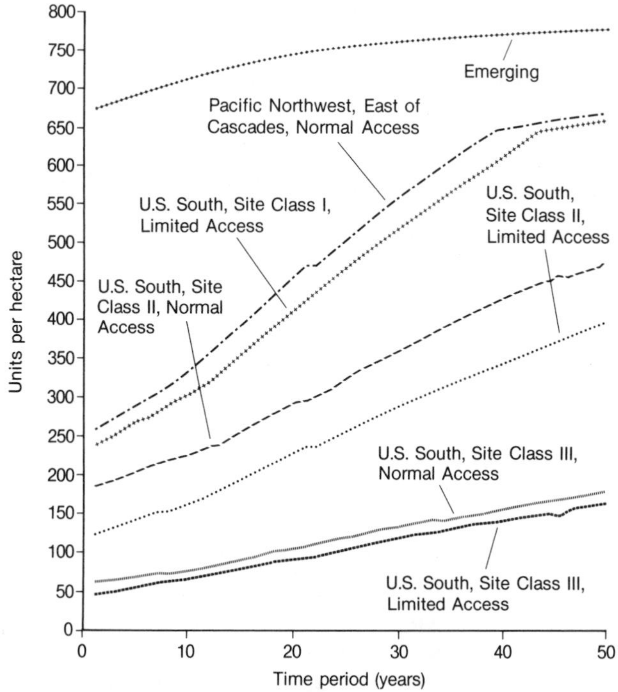

Figure 8-8. Regeneration investment: high-demand scenario.
Note: Investment in the following regions was below 100 units per hectare: British Columbia, Normal Access; British Columbia, Limited Access; Eastern Canada, Boreal/Acadia, Normal Access; Eastern Canada, Boreal/Acadia, Limited Access; Eastern Canada, Lake Forest, Normal Access; Eastern Canada, Lake Forest, Limited Access; Pacific Northwest, East of Cascades, Limited Access; Pacific Northwest, West of Cascades, Normal Access; Pacific Northwest, West of Cascades, Limited Access; Southern Nordic; Northern Nordic; U.S. South, Site Class I, Normal Access; U.S. South, Site Class IV, Normal Access.

high-demand scenario from the previously marginal northern forests of the Nordic countries occur early in the fifty-year period and account for the initial high levels of harvest from the Nordic region (figure 8-5). Similarly, the less accessible regions of Canada have modified the time profile of the harvests for these regions under the high-demand scenario, as shown in figure 8-6.

The effect of high-demand growth is also profound for levels of regeneration investment as projected in figure 8-8. The economically justified levels of investment for each region increase markedly. Although the levels of investment in some of the regions are still quite modest, almost every region and almost every site class can now justify economically investments in regeneration activities on some of its land classes. Regeneration now goes beyond mere reliance on natural processes; high levels of investment in regeneration become economical in much of the U.S. South as well as in the emerging region. For higher-productivity regions the level of regeneration investment becomes substantial.

The Low-Demand Scenario

The low-demand scenario, which posits a worldwide rate of increase in demand of only 0.5 percent annually declining to zero at year fifty, is presented in figures 8-9 through 8-12. The effect of the low growth of demand is manifested in an essentially stable profile of price over time and only very modest increases intertemporally in harvest levels. Harvests of all the world's regions are dampened and levels of investment in regeneration are also reduced, with only the emerging regions and two site classes in the U.S. South justifying significant regeneration investments under an economic criterion. For the low-demand scenario ten of the twenty-two regions do not come into the harvest solution, and this leaves a large residual of forest that is submarginal at the projected prices but that could become available if prices were to increase significantly.

Production from the U.S. South increases much less than in the base case, in part because of the much lower levels of regeneration investment. This leaves a greater market share available for the traditional old growth regions. Nevertheless, these regions are affected by the modest growth in demand. Most seriously affected are western Canada and the Nordic region, both of which are projected to experience quite low harvest levels from fifteen to forty years into the fifty-year projections. However, despite declines to modest harvest levels because of the lower level of demand and price, all the traditional producing regions continue to harvest significant volumes of industrial wood during the projection period.

An interesting feature of the low-demand scenario is that the real price does not decline over the period examined despite the very modest growth in demand. This occurs because the quantity of wood supplied is quite responsive to higher prices. In the absence of strongly increasing demand, therefore, the system simply limits investment in regeneration and chooses to limit severely its expansion into increasing cost areas. Thus the expansion of the quantity supplied is limited merely to eliminating price rises rather than to exhibiting downward pressure on prices. Were biotechnological change to have a greater effect during the period, supply might shift out rapidly enough to elicit a decrease in the real price.

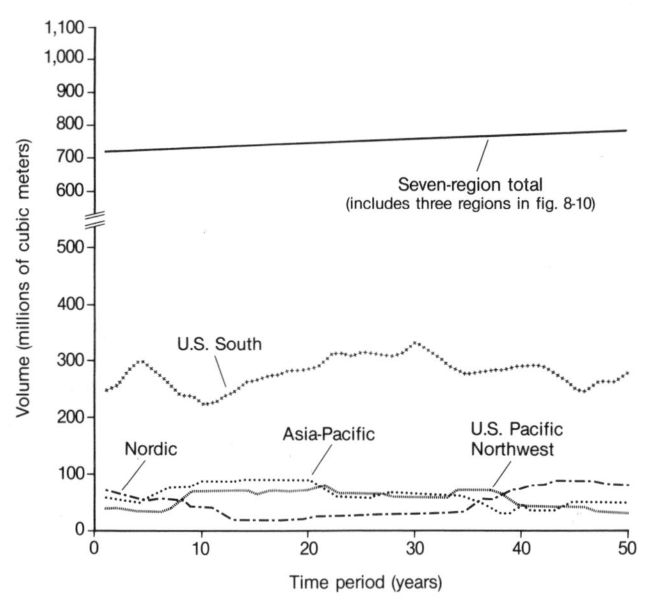

Figure 8-9. Harvest volume over time: low-demand scenario (four regions).

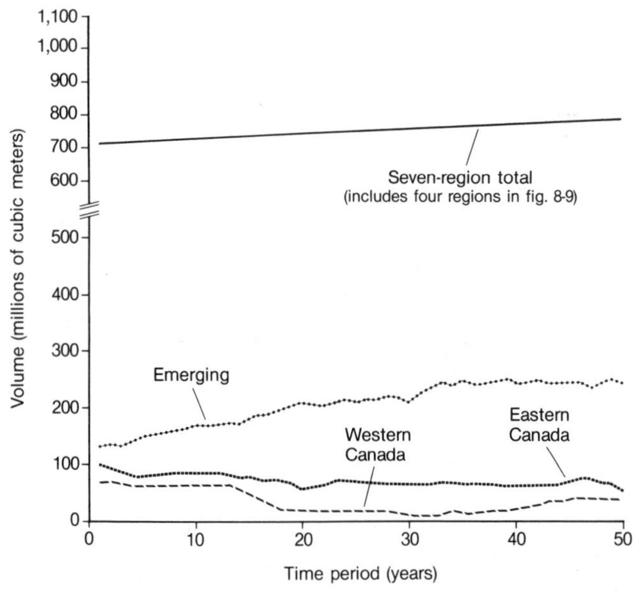

Figure 8-10. Harvest volume over time: low-demand scenario (three regions).

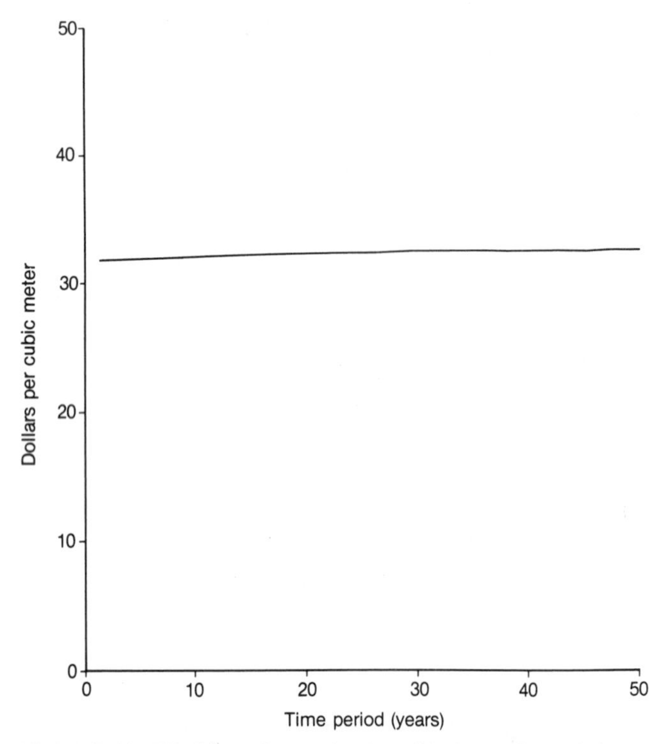

Figure 8-11. World market real price of industrial wood: low-demand scenario.

SUMMARY

The results of these projections are of interest first because they provide a sense of the direction and magnitude of the effects of alternate demand scenarios both globally and for the individual regions. Moreover, scenario analysis can provide some sense of the direction and nature of various adjustment mechanisms. If the model is sound and the empirical base accurate, projections from scenario analysis can provide a sensible empirical estimate of the magnitude of adjustments.

The projections of the base-case forecast, although interesting, are not startling. First, over the fifty-year period from 1985 to 2035, harvests of the entire system are forecast to increase about 30 percent, or about 0.54 percent annually. Second, over that same fifty-year period prices are forecast to rise a modest 12 percent, or an almost negligible 0.2 percent annually. Third, the U.S. South and the emerging region are the only regions forecast to experience appreciable long-term increases in harvests over that period. Production

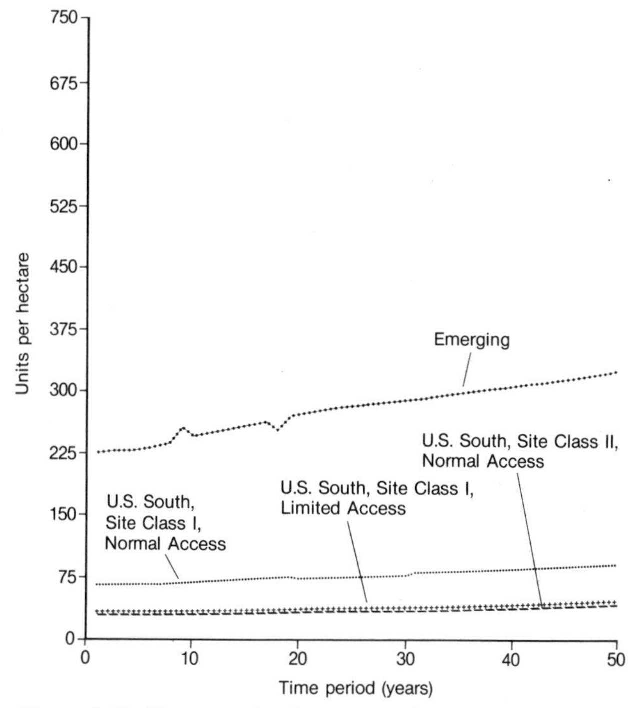

Figure 8-12. Regeneration investment: low-demand scenario.

in the emerging regions rises dramatically to almost 300 million m^3 over the fifty-year period, and harvests in the U.S. South increase about 100 million m^3. These two regions account for essentially all the aggregate output increases experienced by the responsive region.

The traditional producers of northern temperate regions such as Canada, the Pacific Northwest, and the Nordic countries continue to be major producers, albeit in some cases at somewhat reduced production levels, which suggests an erosion of their competitive position. In addition, although some site classes justify relatively high investments in regeneration, many that do not are still projected to continue being important wood harvesters in a worldwide context. Finally, although the overall harvest levels exhibit little volatility, individual regions do experience substantial volatility over the long period examined. This volatility is largest in the traditional old growth timber-producing regions of the temperate Northern Hemisphere.

To the extent that the base case is viewed as the most likely scenario (hence our forecast), the projections reflect an extrapolation of existing and emerging trends and project only relatively modest changes. Harvests and prices are both projected to increase only modestly; harvest projections are within the range of historical experience since the mid-1960s or so, and the price projections are within the growth rates experienced since 1950. Old growth regions experience relative and sometimes absolute declines as timber-producing regions, with their competitive position eroding as the economically accessible old growth becomes increasingly exhausted after the first twenty years. The old growth regions are then forced to adjust to the somewhat lower harvest levels that occur after the transition to second growth. Worldwide, however, increasing production from second-growth forests and the forestry plantations of the U.S. South and the emerging region both offsets the declines in old growth harvests and accommodates the increases required by increasing demand.

The introduction of both higher and lower levels of demand growth provides some sense of the overall sensitivity of the system to changes in these assumptions and also provides insight into how the system might accommodate different demands. Under the high-demand scenario, which assumes that initial demand growth is 2.0 percent annually, gradually falling in successive years to zero after fifty years, harvest volumes and price are projected to rise at a much more rapid yearly rate (0.9 percent and 1.3 percent, respectively), reflecting both the system's ability to expand its production of harvested wood in response to the higher prices and the necessity of higher prices to generate output increases. The system responds to the projected increase in demand growth in a number of different but mutually supporting ways. Price rises, both current and those expected in the future, trigger those responses. Higher current prices allow additional old growth land classes to enter the economic timber base. Higher anticipated prices result in higher levels of investment in regeneration and ultimately in higher levels of harvest.

Structurally, the U.S. South and the emerging region continue to be the largest producers over the fifty-year period. However, some of the old growth regions make substantially higher contributions to harvest levels, as timber stands previously viewed as submarginal by virtue of their location and accessibility are drawn into the economic timber base by the rising real price. In addition, the sources of increased harvests include substantially greater investments in regeneration as the result of incentives created by the higher prices; these investments would be in secondary growth forests as well as in the intensively managed, newly created plantation forests.

The low-demand scenario assumes that worldwide demand will grow initially at a rate of only 0.5 percent annually, declining to zero in fifty years. For this scenario prices show no increase, and the annual growth rate of the worldwide harvest volume is only about 0.25 percent. Investments in regeneration are quite modest, with most regions justifying negligible economic investment levels. Furthermore, more land classes are projected as economically inaccessible for harvesting purposes, with ten of the twenty-two land classes falling into this group.

The regions found to be most responsive to posited changes in demand growth—both increases and decreases—and prices are Canada (both east and west), the Nordic region, and the Pacific Northwest. This is due largely to the fact that these regions have marginal timber lands that become supramarginal as prices rise and submarginal as prices fall. In addition, regeneration investment levels in some regions—for example, the Pacific Northwest—are particularly sensitive to future prices, and the high-demand scenario generates substantial increases in future harvests as a result of investments. These same regions not surprisingly also tend to be susceptible to weak prices. Western Canada in particular is projected to experience relatively low levels of future harvest in the event of low demand growth and relatively weak prices.

In the near term supply can be increased by drawing additional land classes into the harvest. In the long term the level of regeneration investment can have a significant effect on the potential harvest from a land class. This response is readily reversible, with low prices disqualifying regions from consideration for harvest as well as dampening regeneration investment levels.

9

Scenario Analysis

One of the strengths of the type of approach embodied in the Timber Supply Model is the ability to examine an unlimited number of alternative scenarios, that is, simulations that use any number of alternative assumptions about future conditions. The base-case forecast made a number of assumptions that appear sensible but may or may not ultimately be realized. Other sensible assumptions could be made. The scenario analysis, or what is sometimes called sensitivity analysis, allows the base-case assumptions to be modified to determine how sensitive the model simulations are to changes in the assumptions and to generate projections that are consistent with the alternative assumptions. This was done in the previous chapter to examine the implications of alternative assumptions about demand. In addition, scenario analysis allows for the explicit introduction of exogenous "shocks" (external events that might occur at some future time) into the system so that the effects of these shocks on the system's projections can be investigated. The sensitivity analysis can also assist in providing an understanding of how resilient the system may be in adapting to and accommodating various potential shocks.

Thus, although this study is obviously limited in the number of scenarios it can address, using scenario analysis it can examine a wide range of issues—for example, the potential effects of acid rain (pollution), climate change, and major long-term worldwide depression on intertemporal harvest levels, prices, and the regional structure of production. In this chapter we examine four other scenarios and develop their implications. The scenarios are as follows:

1. changes in the structure of world exchange rates;

2. higher and lower levels than in the base case of new forest plantation establishment;

3. a higher harvest level by the Soviet Union than was assumed in the base case; and

4. a change in the tax structure as applied to forestry in one of the major producing countries, the United States.

Three of the scenarios examined here (the first, second, and fourth items in the list) represent a change in the assumptions of the base case that tends to favor (or discourage) production in certain of the responsive regions to the detriment (or benefit) of production in others. In some sense one can think of these scenarios as reflecting changes in comparative advantage among the seven regions formally modeled in this study, and the resulting regional structure of output reflects these changes. For example, changes in the structure of exchange rates generated by circumstances outside the forest sector enhance the competitive position of the currency-depreciating regions at the expense of the appreciating regions. Similarly, the establishment of greater areas of forest plantations in the highly productive and economically competitive emerging region enhances the emerging region's share of total production at the expense of the other regions.

A change in the tax structure of one region that is less favorable to timber production can be viewed as a "shock" external to the system modeled. Although not necessarily reflecting a fundamental change in basic regional comparative advantage, such a policy will induce change in the competitive position of the various regions, thereby creating economic incentives that modify the worldwide structure of timber production. In our example the United States introduces a tax change that inhibits domestic production, thereby enhancing the competitive position and ultimately the production of competing regions.

The fourth scenario, that in which the Soviet Union is assumed to increase production substantially for a thirty-year period, posits a situation in which external events change the market conditions for all seven regions modeled. The effect of these changes, initiated by events outside the system and transmitted into the system via reductions in the level and growth of demand (excess demand) faced by the seven responsive regions, is that prices worldwide and outputs of the responsive regions are affected negatively during the years for which the Soviet Union's production is unusually high. In many respects this scenario has an effect on the formally modeled regions that is similar to that of the decreased demand scenario examined in chapter 8 except that the source of the decreased demand, Soviet wood production flooding the world market, is here clearly identified.

CHANGES IN EXCHANGE RATES

An important factor influencing the structure of worldwide production of timber is the exchange rate or, more accurately, the structure of exchange rates, that relates the exchange value of one currency to another currency. Depreciation of a country's exchange rate will assist that country's competitive position through the supply side by lowering its domestic costs of production vis-à-vis its nondepreciating competitors. In the decade of the 1980s the exchange rate market has been particularly volatile; the U.S. dollar appreciated markedly between 1980 and early 1985 vis-à-vis almost all other currencies and depreciated after early 1985 against most of the major currencies. The exchange rate assumptions used in the base case have been discussed earlier.

In this section we make use of two scenarios—a strong dollar and a weak dollar scenario—to examine the implications for output, price, and particularly the regional output structure of alternative hypothetical exchange rate structures. These alternatives are admittedly hypothetical for several reasons. First, predictions of exchange rate movements are notoriously faulty, and no attempt is made to forecast a "reasonable" time path of exchange rate movements into the future or to introduce such a path into the TSM. Second, it is unlikely that any particular exchange rate level will be maintained for a fifty-year period as is assumed in the scenarios. Third, it is clear that a decline in the dollar does not necessarily imply a decline relative to the currencies of all the responsive regions. For example, the dollar appreciated vis-à-vis many Third World currencies as it simultaneously declined against major world currencies such as the German mark and Japanese yen.

Nevertheless, because there is little question that exchange rates play a major role on the supply side by influencing the cost competitiveness of a region's timber production and that they will probably change over time, it is instructive to examine the way in which alternative exchange rates are likely to affect production and costs and particularly the regional structure of production. Furthermore, because the level and structure of exchange rates may well change significantly over the period examined by this study, an examination of alternative exchange rate structures provides estimates of the long-term implications of different rate levels. The implications of simple alternative strong and weak dollar scenarios for the contributions of the various regions to the structure of worldwide production are projected in the scenarios examining the exchange rate. The three alternative levels of dollar exchange rates assumed in this study are presented in appendix J.

The TSM is particularly well suited to examining the supply-side phenomena, but it is less useful when there are significant feedback effects on demand, as may be the case in exchange rate scenarios. For example, exchange rate changes may generate terms-of-trade effects on domestic demand. Thus a stronger dollar implies increased demand in dollar countries for all normal goods, including wood products, and decreased demand in nondollar countries. The scenario analysis undertaken here assumes that the terms-of-trade effects are completely offsetting in the worldwide market for industrial wood, hence that aggregate dollar demand for timber is unaffected. More generally and perhaps in a more limiting manner, we have made no attempt to adjust aggregate demand to reflect any changes in relative prices that may have been generated by the exchange rate change, and we assume that the dollar demand curve used is invariant to the exchange rate. This is a particular concern when the numeraire, the dollar, is changing systematically its relationship with the other currencies. To the extent that a wide array of countries experience domestic currency changes along with the dollar, this assumption is less distorting. However, to the extent that the dollar change is in isolation, the assumption becomes increasingly unrealistic. It should be noted that the GTM of the International Institute for Applied Systems Analysis (Dykstra and Kallio, 1987) shares the same limitations.

Despite these limitations it is useful to examine the effect of a change in the exchange rate, with particular attention to the effects of the changes on the output of the producing regions in view of the assumption that the aggregate effects on demand are offsetting. Thus the focus of the exchange rate scenario is the effects of a change in the exchange rate structure on the timber supply curves of the various regions.

The Strong Dollar Scenario

The strong or high dollar scenario reflects a situation simulated to reflect roughly an exchange rate such as that which prevailed in early 1985, when the dollar was at the strongest level in recent years (see appendix J). This scenario is presented in figures 9-1 through 9-4. In terms of the time profiles of total harvests and price, the changes from the base case are modest. The regional structure of timber harvests shifts away from the United States, but aggregate worldwide harvest rises slightly as the dollar price declines slightly. The increased harvest is largely due to the markedly higher investments and modest harvest level increases in the emerging region. This is almost but not entirely offset by a reduced U.S. harvest resulting from reduced investments in the United States in response to the modest decline in the dollar price. Hence the aggregate effects offset each other in dollar and nondollar industrial wood markets.

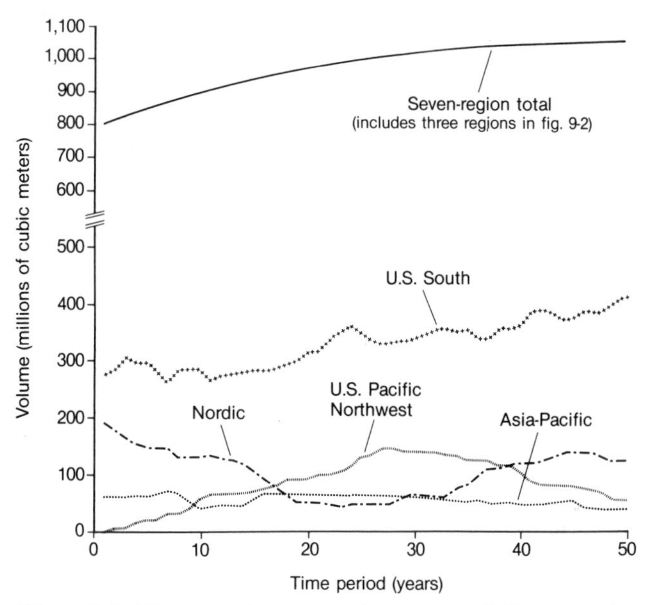

Figure 9-1. Harvest volume over time: strong dollar scenario (four regions).

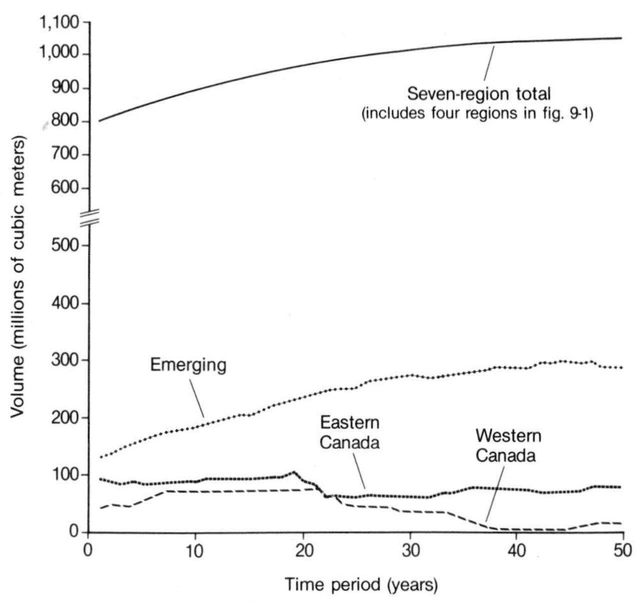

Figure 9-2. Harvest volume over time: strong dollar scenario (three regions).

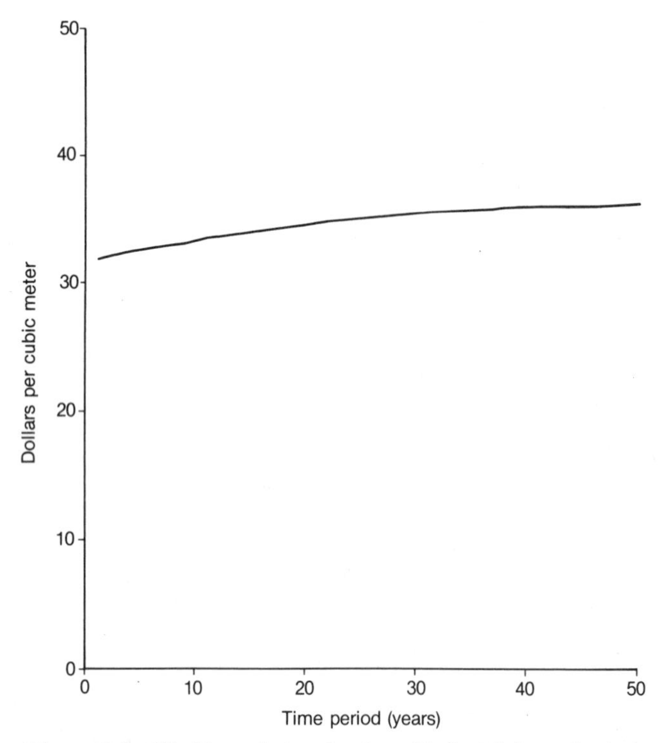

Figure 9-3. World market real price of industrial wood: strong dollar scenario.

The major effect of a changing exchange rate is to change a region's comparative advantage as an industrial wood producer by changing the domestic costs of forest management, regeneration investment, logging, and transportation. For example, a stronger U.S. dollar lowers the dollar costs in non-U.S.-dollar regions relative to the costs of tree growing and harvests in the South and Pacific Northwest regions of the United States.

Because the stronger dollar affects other currencies differentially, the improved competitive position outside the United States will vary. For example, Canada's harvest is changed only minimally, as Canada's currency changes mimic those of the U.S. dollar. Hence, although Canada's competitive position improves slightly relative to that of the United States, this is offset by the decline in Canada's competitive position vis-à-vis that of all the other regions.

More generally, declines in North American harvests reflecting the deleterious effect of the dollar's appreciation on production are offset by increases in harvests in other regions that are now more competitive. Thus the net changes in systemwide harvest and price are negligible, whereas the regional structure undergoes significant adjustments. This is also reflected in

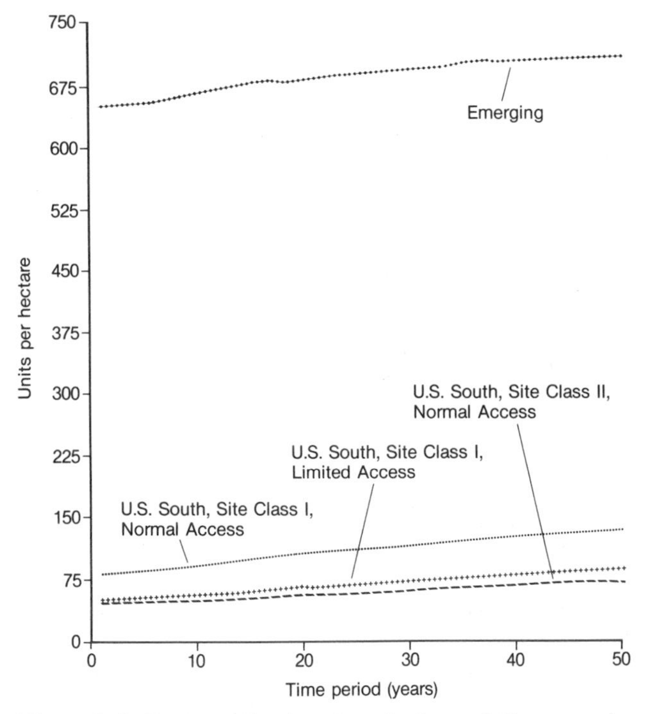

Figure 9-4. Regeneration investment: strong dollar scenario.
Note: Investment in the following regions was below 75 units per hectare: Eastern Canada, Lake Forest, Normal Access; Pacific Northwest, West of Cascades, Normal Access; Southern Nordic; U.S. South, Site Class II, Normal Access; U.S. South, Site Class II, Limited Access; U.S. South, Site Class III, Normal Access.

the projections of regeneration investment in figure 9-4, which shows the shifting of regeneration investments from North America and into competing regions.

The Weak Dollar Scenario

The weak dollar scenario reflects approximately an exchange rate situation such as that prevailing in 1979 and 1980. This scenario is presented in figures 9-5 through 9-8. With a weak dollar the dollar price time path is higher than under the base case, and the aggregate harvest level is lower. Regionally, U.S. production increases absolutely and relatively, responding to both its improved competitive position and the higher dollar price. Simultaneously, foreign production in nondollar countries declines, reflecting both the higher dollar cost equivalents of investments in regeneration and logging and access costs, particularly in higher-cost and less accessible regions. The

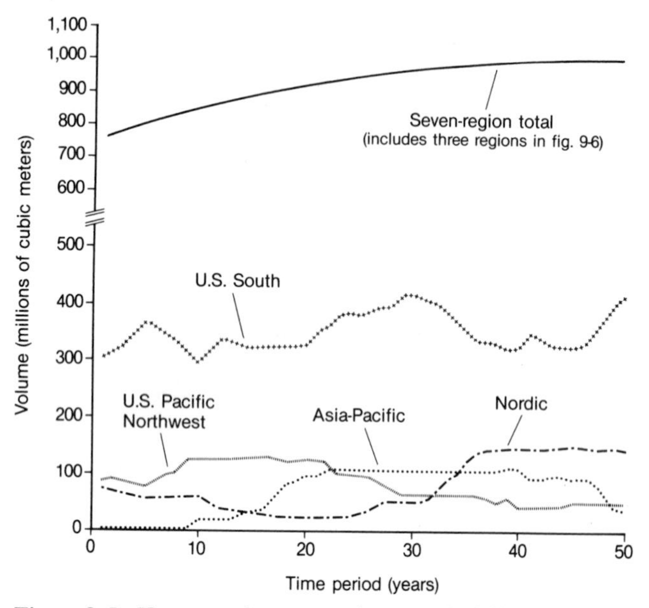

Figure 9-5. Harvest volume over time: weak dollar scenario (four regions).

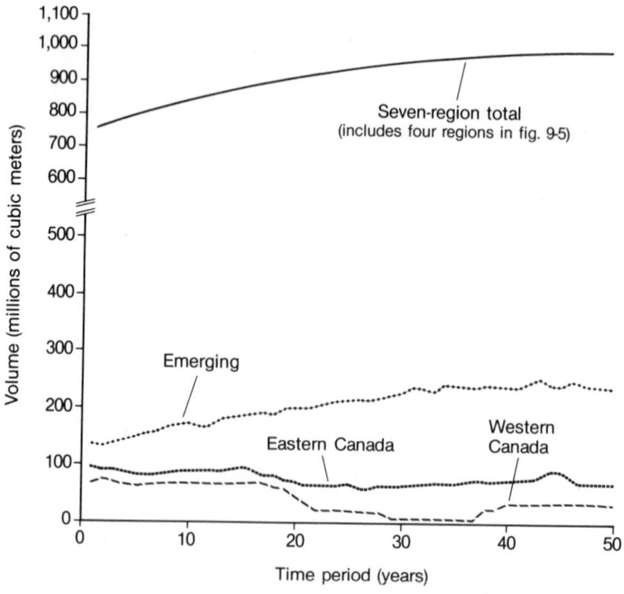

Figure 9-6. Harvest volume over time: weak dollar scenario (three regions).

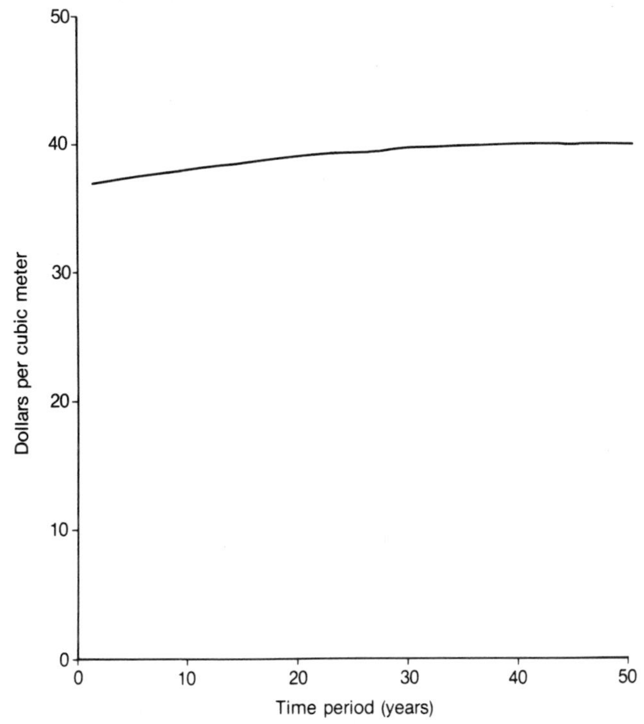

Figure 9-7. World market real price of industrial wood: weak dollar scenario.

exchange rate effect of the weak dollar is particularly significant for the non-North American regions, where the exchange rate change is assumed to be large. The decline in harvests in these regions from their deterioration in competitive positions is offset somewhat by the higher dollar price, but this offset is small and incomplete.

The weaker dollar gives U.S. producers a competitive advantage over producers in nondollar countries that is reflected in higher harvest levels. Because Canada's exchange rate is assumed in this scenario largely to mimic that of the United States, Canada also shares an advantage vis-à-vis non-North American producers. By contrast, non-North American regions' production levels are modified and reduced significantly. The emerging regions experience a reduction of investment in regeneration, timber growth, and harvests. The changing structure of costs also influences the time at which some regions begin production for the world market. For example, the Asia-Pacific region delays significant harvesting for almost a full decade, and the northern land class of the Nordic region is also delayed dramatically in its entry as an economic region.

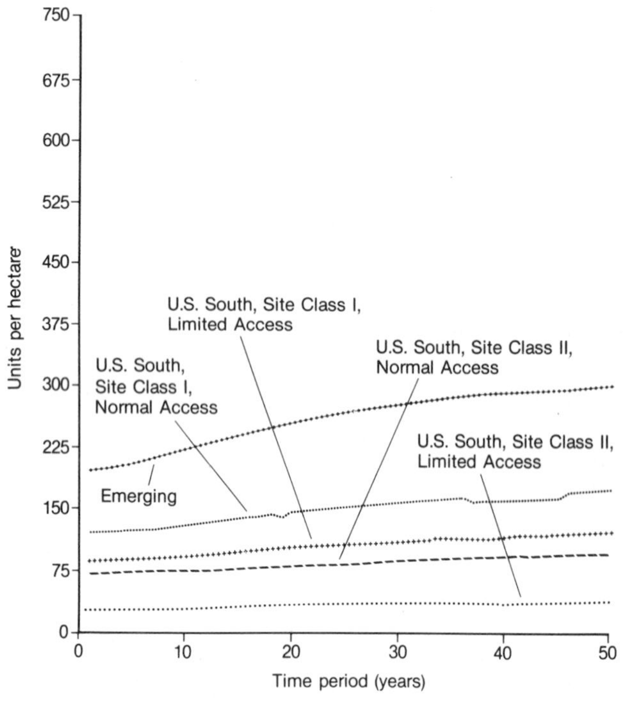

Figure 9-8. Regeneration investment: weak dollar scenario.
Note: Investment in the following regions was below 50 units per hectare: Eastern Canada, Lake Forest, Normal Access; Pacific Northwest, West of Cascades, Normal Access; Pacific Northwest, West of Cascades, Limited Access; U.S. South, Site Class III, Normal Access.

An Additional Effect on Processing Costs

It might be noted that although this analysis suggests that changes in the exchange value of the dollar will affect investments in and production of industrial wood in a region and thereby modify the overall structure of timber production worldwide, for exchange rate changes of even the substantial levels that have been experienced in recent years and examined here, the changes in overall structure of industrial wood production have been shown to be modest. However, the effect on the forest products industry as a whole will almost surely be considerably larger than for the production of raw wood delivered to the mill, which is the focus of this study. This is because exchange rate changes will affect comparative domestic processing costs as well as delivered-wood costs, hence the competitive position of the various wood-processing activities. Thus, for example, the effect of a rising dollar on the U.S. forest products industry as a whole will be felt through both the decrease in timber-producing activities resulting from higher relative delivered-wood costs as well as decreased domestic processing activities resulting from higher relative domestic processing costs.

A Comparison of Scenarios

Although simulations must be interpreted carefully because of the limiting assumptions regarding the effect of exchange rate changes on aggregate demand, they demonstrate that the exchange rate can help or hinder the competitive position of the various regions as producers of industrial wood for a given dollar level of aggregate demand. This response to the exchange rate comes in the form of changes in both the levels of investment in regeneration and in the timing and extent to which marginal forests participate as producing forests.

A somewhat surprising result is the lack of symmetry between the strong and weak dollar scenarios. The strong dollar scenario creates little in the way of aggregate changes in price or harvest; in contrast, the weak dollar scenario does generate significant changes. The explanation for this asymmetry appears to be the greater sensitivity of investment and harvests in the ranges examined to the higher dollar costs faced by the nondollar countries for both regeneration and the operating costs of harvests. A strong dollar, implying lower foreign investment costs per unit of output, engenders a larger increase in harvest, and a weak dollar discourages foreign production far less.

THE ESTABLISHMENT OF NEW FOREST PLANTATIONS

One of the major uncertainties in attempts to project future timber harvest levels relates to the extent to which new industrial forest plantations utilizing exotic (nonindigenous) species in the emerging region continue to be established and subsequently become major sources of industrial wood. These plantations are located largely in the Southern Hemisphere semitropical regions of such countries as Brazil, Chile, Venezuela, New Zealand, Australia, and South Africa as well as Spain and Portugal in Europe. In the base case it was assumed that the existing industrial plantations area of about 6 million hectares in the emerging region will be harvested when the trees reach maturity. In addition, the base case assumes that new plantations will be established at the rate of 200,000 hectares per year for thirty more years to a total of about 12 million hectares of high-yielding plantation forest.

This scenario examines the implications of both higher and lower levels of plantation establishment within the emerging region. It is widely accepted

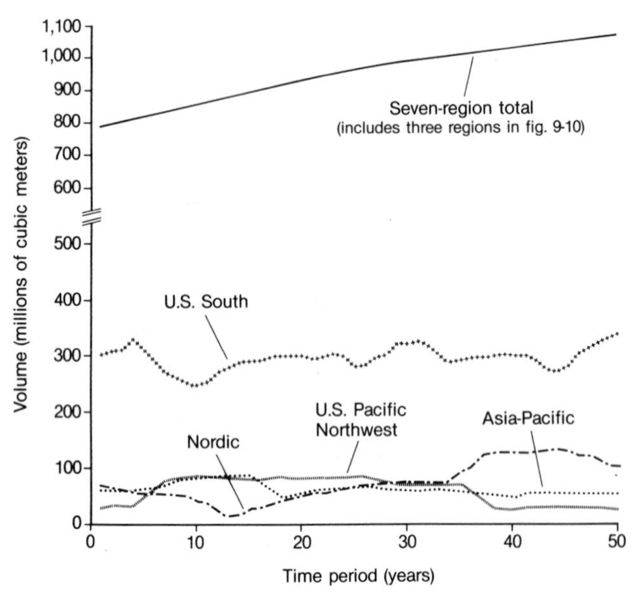

Figure 9-9. Harvest volume over time: high plantation starts (four regions).

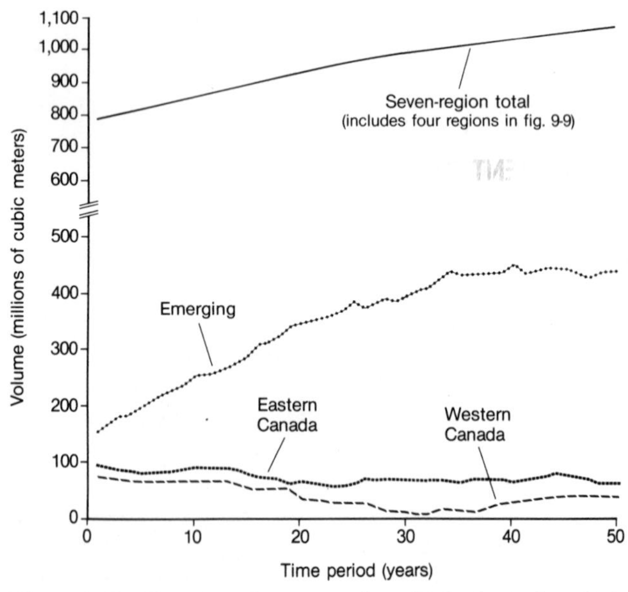

Figure 9-10. Harvest volume over time: high plantation starts (three regions).

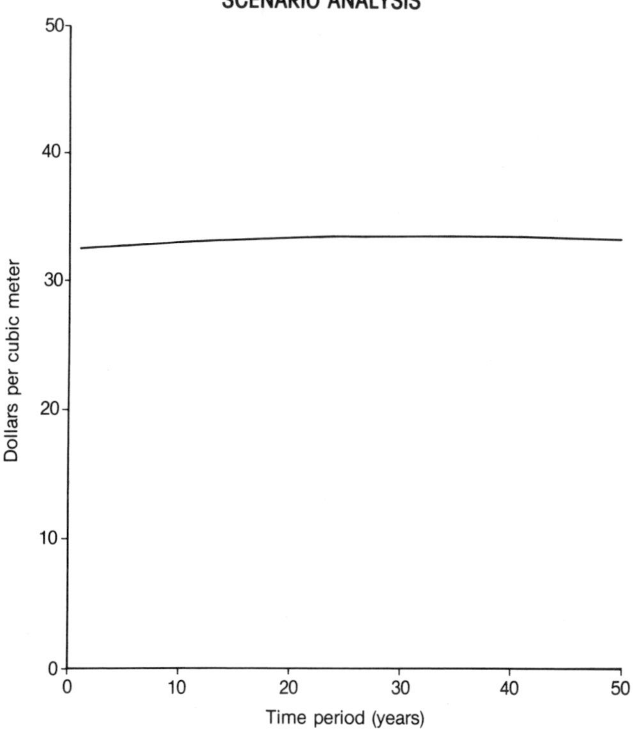

Figure 9-11. World market real price of industrial wood: high plantation starts.

that the countries identified within the emerging region have the potential to accommodate the range of levels of new plantings considered in this study.

Scenario of High Plantation Starts

The scenario of high plantation starts assumes the same initial area in plantations as in the base case but further assumes that the rate of plantation establishment will be 500,000 hectares per year for thirty years for a total of 21 million hectares. This is approximately the rate of new plantation establishment that prevailed in the latter half of the 1970s. The projections appear in figures 9-9 through 9-12.

The effect of high plantation starts is to produce higher regional harvests in the emerging region than were found in the base-case forecast. This results in a modest lowering of the world market real price for industrial wood under the high plantation start scenario (figure 9-11), which in turn somewhat discourages regeneration investments in other regions. Thus the higher plantation starts in the emerging region are offset, but only in part, by lower regeneration investments elsewhere. Overall, total net harvest is higher than in the base-case forecast, and the world market real price is slightly lower.

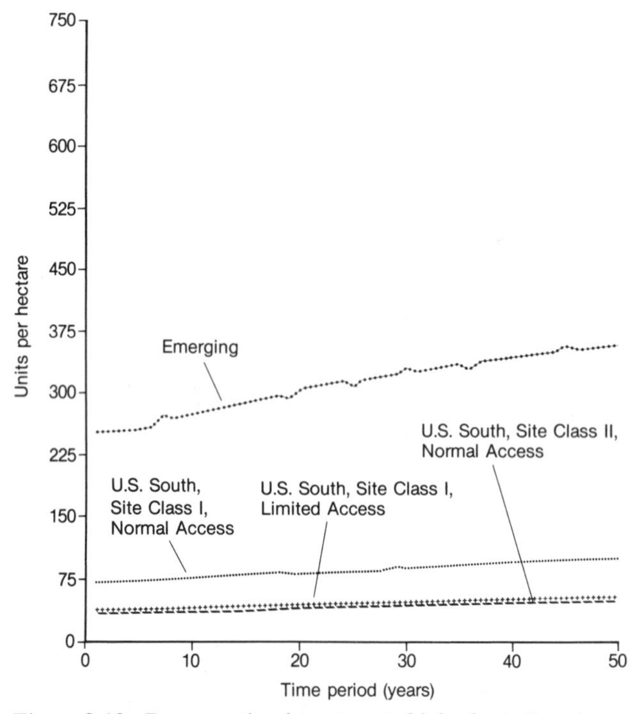

Figure 9-12. Regeneration investment: high plantation starts.

Major changes are found in the structure of the regional harvests, with investment in regeneration dampened throughout the world. For example, unlike some of our earlier base-case forecasts, harvests in the U.S. South do not rise significantly over the fifty-year period. Rather, although harvests in the U.S. South are similar in the initial years to those of the base case, investments in regeneration are significantly lower, and the long-term increases in the harvests of the U.S. South projected in the base case never transpire in the high plantation scenario. By the end of the fifty-year period, the emerging region has displaced the U.S. South as the dominant timber-producing region. In addition, the higher volumes harvested from the emerging region displace some production from the other supply regions as well and generally modify the time profile of these regions' harvests.

Scenario of Low Plantation Starts

The scenario of low plantation starts assumes no new plantations after 1985, thereby fixing 6 million hectares in industrial plantation forests. The effect of low plantation starts is, not surprisingly, just the opposite of that of the high start scenario. A modest decline in total harvests vis-à-vis the base-case forecast and a greater increase in price is shown in figures 9-13 through 9-16. In this scenario prices rise about 20 percent over the first fifty years—about twice the increase of the base-case forecast. The higher prices induce higher levels of investment in regeneration in the limited forest area of the emerging region as well as elsewhere. The result is increased harvests from the other regions and a structure of worldwide production in which the emerging region plays a relatively modest role.

The U.S. South is affected strongly by this action; higher market prices result, inducing an increase in the level of regeneration investments and subsequently large increases in harvests. Other regions also experience higher levels of regeneration investment; however, a major source of their increased harvest levels is found in the marginal forest regions that are drawn into the timber base by the higher real prices occurring in this scenario.

A Comparison of Scenarios

The difference in the systemwide, long-term harvest of high plantation versus low plantation establishment is relatively modest. The high establishment scenario ultimately generates annual systemwide harvests that are only about 5 percent or 50 million m^3 higher after fifty years than in the low establishment case. The price effect, however, is about 20 percent higher for the low establishment case. The regional structure of harvests is dramatically affected by the assumptions regarding plantations. The effect on the emerging region in particular is quite profound, with harvests from this region about 300 million m^3 lower in the low establishment case as compared with the high establishment case.

The difference between the large decline in the harvests of the emerging region and the very modest decline in total harvest is accounted for by increases in the harvests of the other regions in response to the higher real prices they face. The systemwide price rises substantially and elicits a level of regeneration investment that offsets much of the harvest decline in the emerging region. The amount that can be invested economically in the relatively limited available forestland of the emerging region is bounded by decreasing returns to scale and gives rise to higher levels of regeneration in the other regions—about twice the levels of investment in regeneration in most of the regions as in the high establishment case. (This can be seen by comparing figure 9-16 with figure 9-12.) The projections also demonstrate the role of the submarginal forests that could be drawn into the economic timber base if timber prices were to rise, as would be the case were plantations from the emerging region not developed into a major supply source.

The results of this simulation demonstrate the dominant role that industrial forest plantations can play in meeting the future demand for timber. The

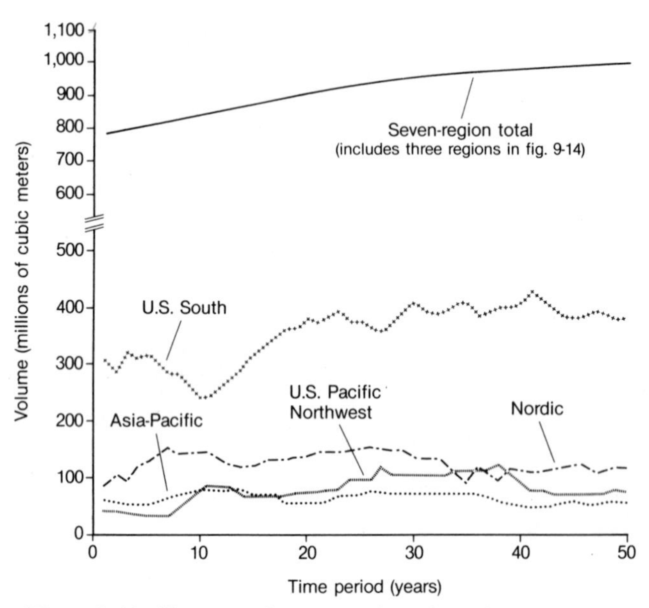

Figure 9-13. Harvest volume over time: low plantation starts (four regions).

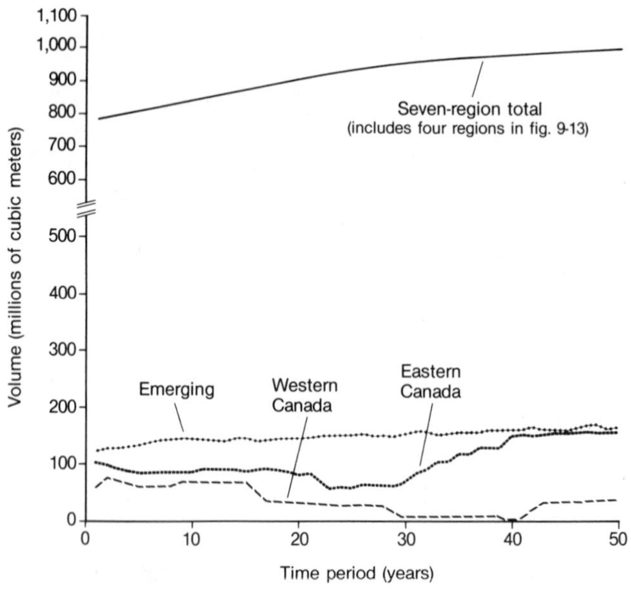

Figure 9-14. Harvest volume over time: low plantation starts (three regions).

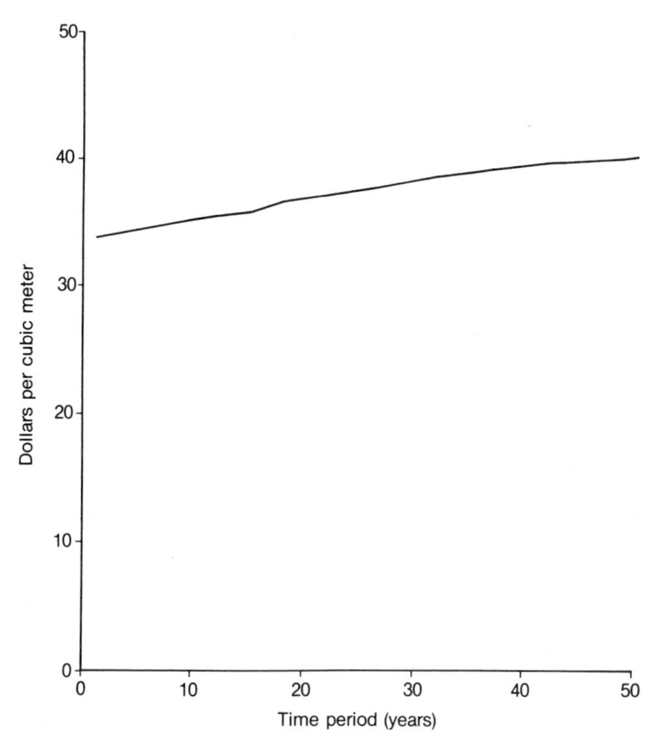

Figure 9-15. World market real price of industrial wood: low plantation starts.

simulations suggest that the emerging region could become the largest single industrial wood-producing region among the regions represented in the system. Our scenario analysis also reveals the important interrelationships among the various regions. For example, the future role of the U.S. South depends importantly on the level of establishment of new forest plantations in the emerging region, something over which the U.S. South has no control. Finally, the scenarios also demonstrate the resiliency of the system and its lack of dependence on the harvest of any single region. Should one region fail to assess properly its opportunities to establish (or regenerate) its forests, price rises will induce other regions to move to fill in any potential void.

INCREASING SUPPLY FROM THE SOVIET UNION

Although the TSM focuses attention on production and investment in the responsive regions, this must not be interpreted as suggesting that changes in regions not responsive to market forces have little effect on world markets.

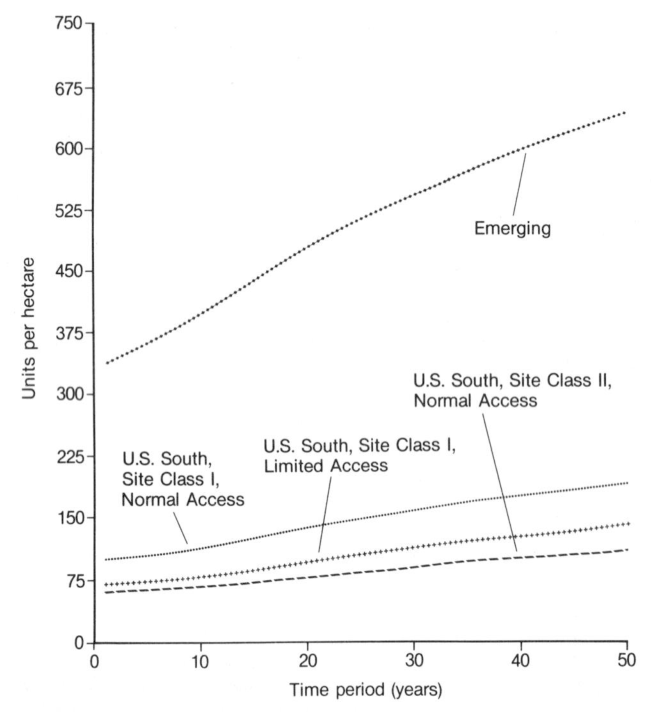

Figure 9-16. Regeneration investment: low plantation starts.
Note: Investment in the following regions was below 50 units per hectare: Eastern Canada, Boreal/Acadia, Normal Access; Eastern Canada, Lake Forest, Normal Access; Pacific Northwest, West of Cascades, Normal Access; Pacific Northwest, West of Cascades, Limited Access; Pacific Northwest, East of Cascades, Normal Access; Southern Nordic; U.S. South, Site Class II, Limited Access; U.S. South, Site Class III, Normal Access; U.S. South, Site Class III, Limited Access.

Rather, the effect of actions by the nonresponsive countries on markets can be profound and in essence require the market-driven regions to adjust and adapt to changes initiated by non-market-driven forces and decisions.

As of the mid-1980s the responsive and nonresponsive regions identified in this study divided the world's total production of industrial wood harvest almost identically between them. Among the nonresponsive countries the most important wood producer is the Soviet Union, in terms both of current production and also of a vast inventory for timber that is potentially available for exploitation.

The base-case forecast assumes that the harvests of the nonresponsive regions grow autonomously at an annual rate of 0.5 percent, with the growth rate gradually falling in successive years to zero after fifty years. The burden of providing the rest of the production to meet total world requirements then falls on the responsive regions, whose production is assumed to respond to economic forces. The demand facing the responsive regions can be viewed as total world demand less that met by the nonresponsive regions, or what is called residual or excess demand. In this context, for a given worldwide demand, increased supply from the nonresponsive regions will diminish the excess demand facing the responsive countries. Hence changes in the level of supply generated by the nonresponsive regions will have an effect on the level of demand facing the responsive regions.

The following scenario examines the effect of autonomous increases in the production of the nonresponsive regions. Although the increase is assumed to be the result of harvest levels in the Soviet Union that are higher than were assumed in the base-case forecast, the increase could originate in any country or group of countries in the nonresponsive set. In this scenario we assume that the harvests of the Soviet Union experience an increase of 50 percent over a twenty-year period and remain at this higher level indefinitely. Specifically, beginning in year ten (1995) through year thirty (2015), the production of the Soviet Union (hence of the entire nonresponsive region) increases cumulatively at 12.5 million m^3 per year more than is assumed in the base-case forecast. The increase remains part of the output indefinitely. The effect is transmitted to the TSM and the responsive regions through a downward shift of the appropriate amount in the (excess) demand curve facing the seven responsive regions.

The projections of this simulation are presented in figures 9-17 through 9-20. For the high Soviet harvest scenario, both the world real price and the production of the responsive regions show a great deal of stability over time. The world market real price rises less than in the base case—only a few percentage points—and is essentially stable once the twenty-year increase in the Soviet harvests begins. As a consequence harvests in the responsive regions show only modest increases, most of them occurring before the increases in Soviet harvests begin. The relatively low wood price induces only modest investments in regeneration in the various responsive regions and is subsequently translated into only modest regional harvests; few areas experience long-term overall growth.

The results of this scenario indicate the dampening effect that large increases in Soviet harvests, or harvests in general from any nonresponsive country or group of nations, are likely to have on the global market for industrial wood as compared to the base case. The dampening in price results in a reduction in regional regeneration investments, which in turn leads to lower future harvests for the responsive regions. The larger the harvest of the nonresponsive regions and the less (excess) demand faced by the responsive markets, the lower world prices and the less market forces create an incentive for increased harvests.

Furthermore, the results of this scenario emphasize the importance of the interrelationship between harvests in nonresponsive regions and investments and harvests in the responsive regions. The production of the two regions,

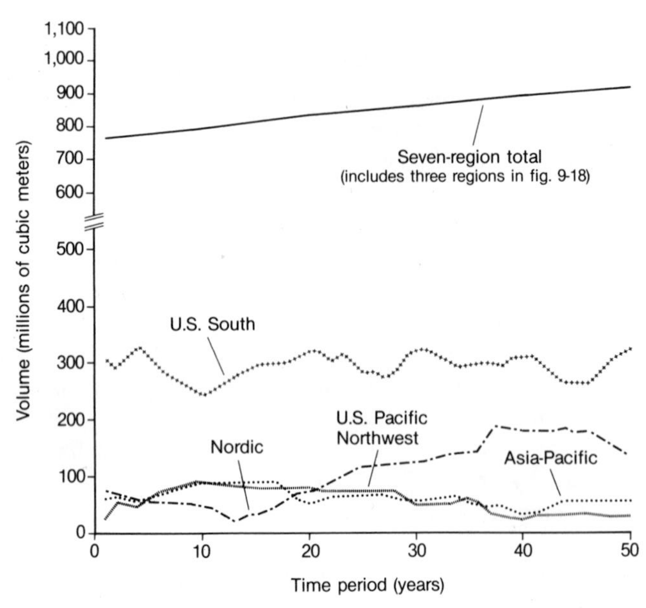

Figure 9-17. Harvest volume over time: Soviet case (four regions).

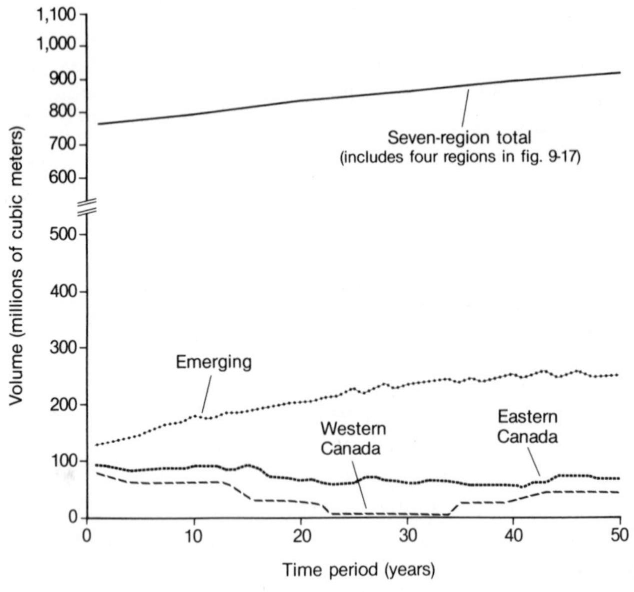

Figure 9-18. Harvest volume over time: Soviet case (three regions).

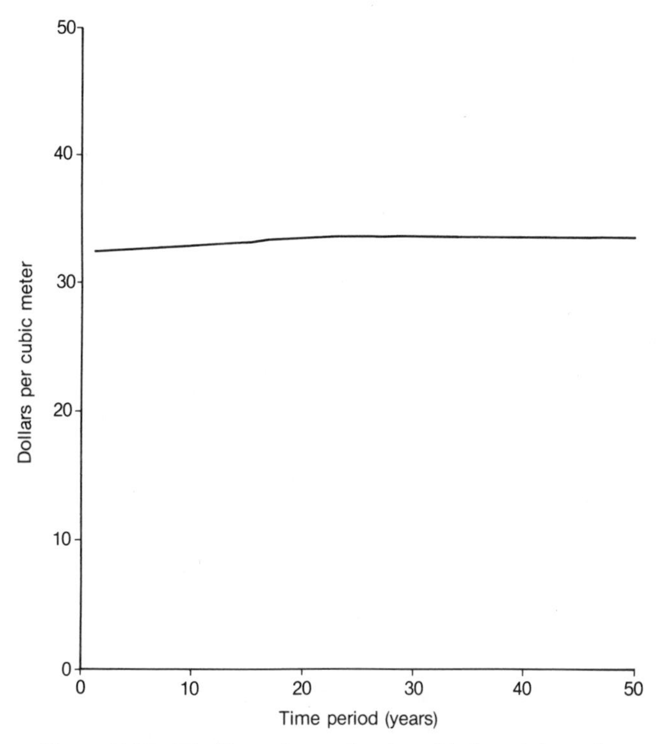

Figure 9-19. World market real price of industrial wood: Soviet case.

one market driven and one autonomously determined, are highly interrelated, as ultimately they service the same world market. The market-driven regions adapt to the changes initiated elsewhere and serve to dampen the instability in world markets resulting from forces operating outside the market system.

A MAJOR TIMBER TAX INCREASE IN A MAJOR REGION

We now turn to the effects on the system (both regionally and worldwide) of a major tax increase on investments in timber regeneration in a major producing country. The example we will examine is that of the effect of a hypothetical major change in U.S. tax laws that would reduce the tax benefits associated with investments in regeneration. Within the context of this study, the purpose of this scenario is to examine the international ramifications of an ostensibly domestic policy not generally perceived as having significant international impacts. As is demonstrated in this scenario, however, anything

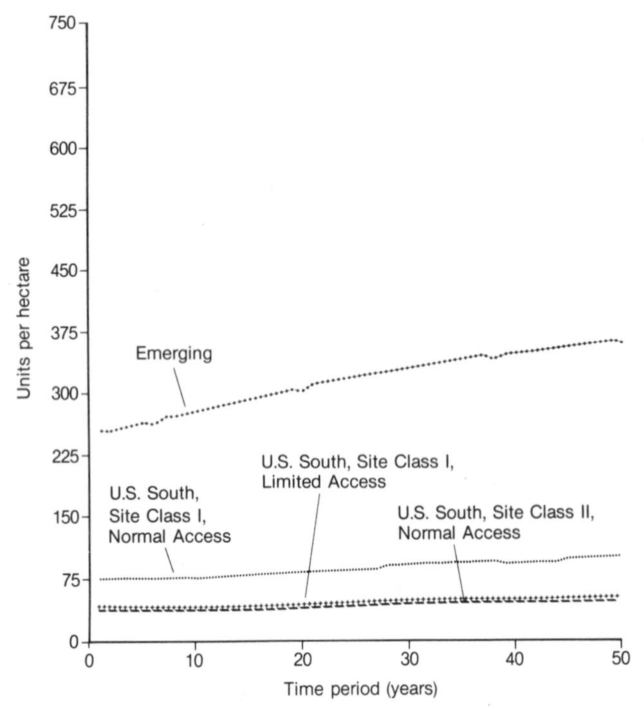

Figure 9-20. Regeneration investment: Soviet case.

that has a major effect on the production of a major producing region will also have important consequences beyond that region.

This section assumes that tax changes are introduced that initially reduce the internal rate of return by 40 percent (Sedjo, Radcliffe, and Lyon, 1986). This change is then introduced into the TSM as a shift of the investment function for the affected regions, the U.S. South and the Pacific Northwest. The modification, which lowers the optimum levels of investment in regeneration and ultimately lowers long-term regional harvests in the United States, captures the long-term effect of the tax change on the levels of investment in regeneration in the United States. Changes in the regional harvest levels result in further adjustments throughout the worldwide timber-producing system.

The projections of major changes in the U.S. timber tax, together with the original base-case forecast, are presented in figures 9-21 through 9-23. The base-case forecast, which implicitly incorporates the current U.S. timber tax system, is compared with projections under the hypothetical new higher tax. Figure 9-21 presents the long-term intertemporal time path of real price under the two alternative assumptions regarding the timber tax. As can be seen, under the higher timber tax the time path of the world market real price is higher for each time period. The price is higher even in the initial period, as producers in all regions realize that future supplies will be more scarce and therefore postpone some harvesting decisions, awaiting higher prices. The effect is to drive up the prices well in advance of the effect that decreased investment has on future timber growth. However, even with a long-term world price rise, price increases in both the short and the long term are only a modest 8 percent above the base-case forecast.

Figure 9-22 presents harvest levels over time for both the world system and the United States. The projections show that although the long-term decline in worldwide harvest is only a modest 24 million m^3, the decline in U.S. harvests is substantial, converging to a reduction of about 51 million m^3, or about 12 percent of U.S. production in the long term. The absolute decline in the U.S. harvest is almost twice the decline in total world harvests because competing foreign regions expand their investments and harvests in the face of reduced domestic U.S. investments and higher current and anticipated future world market prices. The rise in price is limited to only about

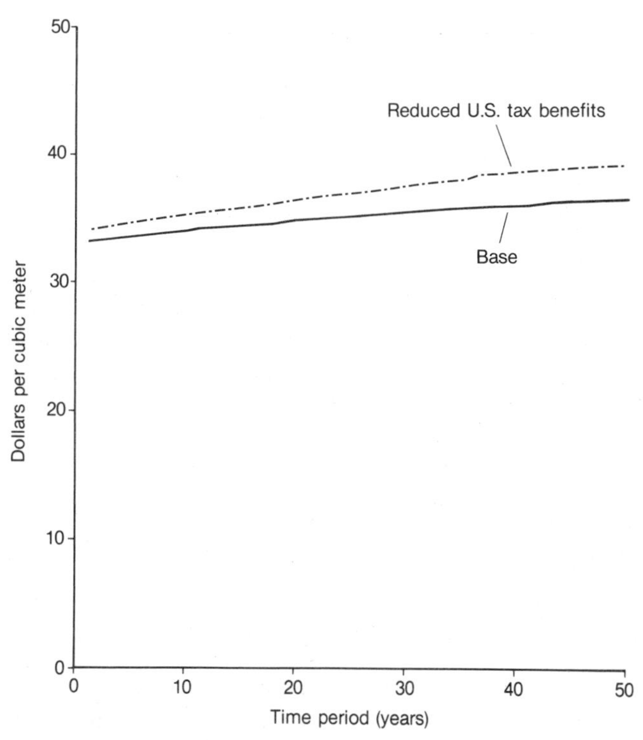

Figure 9-21. Real price of wood over time.

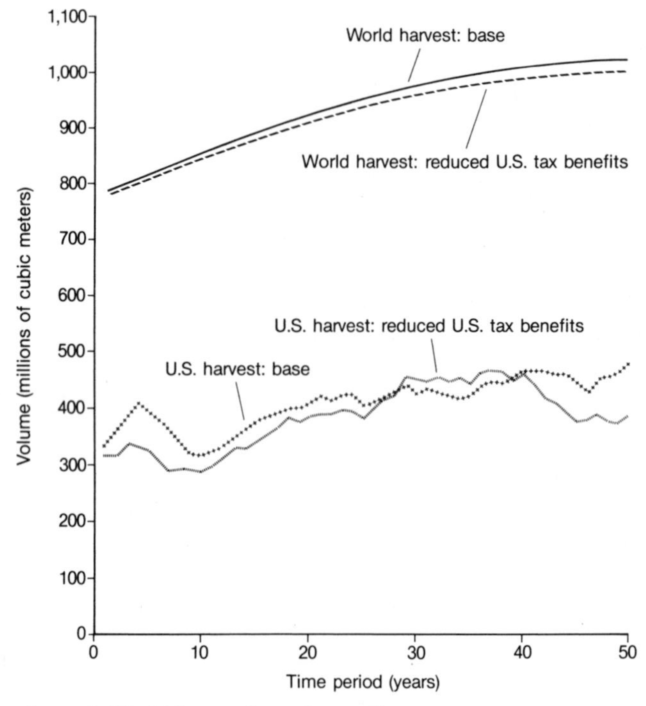

Figure 9-22. Volume of wood over time.

8 percent because the increasing non-U.S. production, which offsets much of the harvest decline created by U.S. reductions, also moderates the price increase.

The effect within the United States is moderated because the system is open to international trade. American consumers experience only modest increases in prices in the face of large declines in U.S. production. However, not only are American producers forced to reduce harvests in the face of the higher taxes, but also the moderation of price rises, due to trade, dampens the beneficial effect to producers of the higher prices. In a system insulated from the rest of the world, the price rise would have to be greater, thereby allowing producers to offset some of the tax-generated revenue losses with revenues generated by higher prices. However, because the system is open to international trade, this effect is reduced significantly.

The Global Implications of a Tax Increase

In light of interregional substitution, individual regions are not insulated from occurrences elsewhere in the system. Adjustment occurs through redirections in investment and consequent relocations of production. As figure 9-23 shows, projections of intertemporal investment in the United States

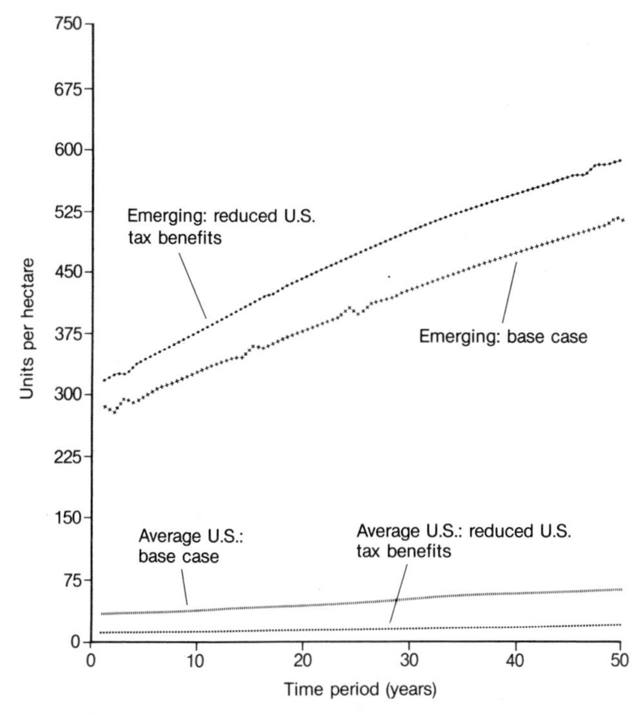

Figure 9-23. Regeneration investment over time.
Note: Investment in other regions was negligible.

decline as investments in other regions, particularly the emerging region, rise. In a global context the scenario of a tax increase again demonstrates the high degree of regional interrelatedness and the potential for a high degree of interregional substitution in regeneration investments and harvests. Declines in regeneration investments and harvests in one region can set in motion forces that will significantly offset these declines by inducing increased investments and harvests from other regions.

Supplying regions cannot be viewed in isolation but must be viewed as an integral part of a systemic whole. What appear to be domestic policy decisions affecting a single political entity in fact have broad ramifications for all the producing regions.

CONCLUSIONS AND QUALIFICATIONS

This chapter examines several alternative scenarios to the timber supply base-case presented in chapter 8. We have introduced changes in the base-case assumptions of the TSM to simulate the effect on the system of these changes. Because the model is addressing long-term considerations and abstracts from the business cycle, it should be recognized that the interrelatedness is also long-term in nature. The focus of this analysis is on changes in total harvest and prices from the base-case results and on the change in the regional structure of investments and outputs. Different types of scenarios are examined, including both external and internal "shocks" in the form of an increase in Soviet timber harvest and an increase in the tax on U.S. production. In addition, the effects of changes in several important assumptions regarding relevant but essentially unknowable future events—for example, exchange rate changes and the level of new plantation establishment—are examined.

Although we have investigated only a small number of alternative scenarios, it should be clear that the possibilities are almost endless and that a wide variety of situations and potential events can be examined using scenario analysis. The examples used also point to the limits of useful simulations. For example, when a particular change affecting the supply side may have substantial effects on the aggregate level of demand that cannot be estimated easily, as in the exchange rate scenario, the usefulness of the simulations declines. This feature emphasizes the essential partial-equilibrium nature of the TSM.

The scenarios of this chapter, together with the investigations of alternative demand levels in chapter 8, lead to the following two general observations. First, there is a high degree of interrelatedness between production in one region and that of another. Changes in harvest and/or regeneration in one region are partially offset by changes in the opposite direction in other regions. Similarly, a shock affecting supply in one region induces actions in other regions that partly mitigate the global effect of the shock. The driving mechanism is price, which signals the producers of all responsive regions as to the relative availability (or lack thereof) of timber, and to which producers in responsive regions adjust their level of investment in forest management and regeneration. In essence, diversification of production across regions also diversifies the global effect. This result ought not to be viewed as surprising when one recognizes that the regions all produce for what is essentially consumption in a common world market, hence all respond to a common world price (or better, structure of prices).

Second, although the effect of regional shocks on global supply can be significantly offset by a systemwide response, the strains on the system that are generated when demand is projected to grow at a fairly rapid rate for an extended period of time are not accommodated as easily. For example, the high-demand scenario clearly outran the ability of supply from all sources to increase harvests, and the result was rather sharp increases in the rate of increase of harvest real prices. Hence, certain global stresses may require a systemwide response that involves more than simply interregional adjustments to mitigate the effects. For example, the systemwide effect of an increase in the rate of growth of global demand (as discussed in chapter 8) requires a continuing expansion of harvests (relative to what it would have been) in all regions.

The TSM has two mechanisms to deal with a situation of sharply increasing total worldwide demand. The first mechanism is part of traditional economics and works through price rises, which induce increased production within the constraints of the relatively static production potential and simultaneously choke off demand. The second mechanism in the system that could deal with increasing demand is technology. Changing technology is introduced into the TSM; it both moderates the rate of growth in demand (e.g., through wood-saving innovation) and affects timber supply through technology's effect on timber yields and utilization of previously unwanted species.

In the TSM the effect of technology on supply is specified to be autonomous (unrelated to factors in the model) and to have only a modest quantitative effect. In the real world technology may be much more powerful and dynamic. For example, real-world technological change is almost certainly induced at least in part by higher prices. Hence it is possible, and perhaps even probable, that a more rapid rate of demand growth than is posited in the base case would trigger more rapid introduction of technology in forestry than is incorporated in the TSM and thereby offset more of the demand-induced pressures than is suggested by the model. In the very long term the direction of real prices will depend on the outcome of a race between consumption growth and technology. If, as the base-case forecast projects, consumption growth is not too rapid or if technology develops more rapidly than has been posited, a high degree of long-term stability in real wood prices would be expected.

REFERENCES

Dykstra, Dennis P., and Markku Kallio. 1987. "Scenario Variations," in M. Kallio, D. P. Dykstra, and C. S. Binkley, eds., *The Global Forest Sector: An Analytical Perspective* (New York, Wiley).

Sedjo, Roger A., Samuel J. Radcliffe, and Kenneth S. Lyon. 1986. "Tax Reform and Timber Supply." Policy Brief (Washington, D.C., Resources for the Future).

10

Conclusions and Implications

This chapter summarizes the approach and findings of this study. We began by presenting a brief summary of the study's objectives and method. We then summarize the major modeled projections of the long-term behavior of real prices and timber harvests and compare these with projections from other recent studies; in addition, we discuss some of the implications of this comparison. Then we draw some general conclusions regarding long-term forest resource availability based on the findings of the study. Finally, we summarize some of the study's major findings.

THE STUDY'S OBJECTIVES AND METHOD

The major objective of this study is to investigate the question of the long-term worldwide availability of industrial wood. A secondary objective is methodological—the development of an optimal control timber supply model that cannot only assist in the investigation of timber supply but that also has broader applicability because it provides a technique that can be used to examine a host of questions related to forestry and natural resources. To this end the study has presented the following:

1. a brief history and background of resource and timber supply scarcity issues;
2. an overview of the world's forests resources;
3. the development of the TSM in both intuitive and technical versions;

4. the role of technological change and its long-term effects on supply and demand;
5. the development of the TSM's demand-side assumptions;
6. a base-case forecast;
7. several alternative scenarios; and
8. the study's principal implications.

The methodological approach of this study has been to use the TSM as the major vehicle for examining the implications of analyses of important determinants of future availability, such as growth in demand and technology. The formal model is supplemented by two other separate but related analyses: the analysis of the long-term demand for industrial wood (primarily in chapter 3) and that of the effect of technological change on intertemporal timber availability (in chapter 6).

The implications of the analyses of demand and of technological change for long-term timber supply are further developed by introducing formally certain elements of these considerations into the model. For example, the demand assumptions used in the model are developed largely on the basis of changing historical trends and are then introduced into the model. Similarly, some of the effects of technological change are introduced (albeit only imperfectly) into the model.

Using the completed TSM, we then make a series of projections to simulate the implications of specific assumptions and external "shocks" on harvest levels and prices over fifty years (1985 to 2035). Such an approach allows us to integrate the analysis of historical consumption trends and recent technological innovations with supply considerations such as biological growth rates, locational considerations, harvest, and related costs.

AGGREGATE TSM RESULTS AND A COMPARISON WITH OTHER PROJECTIONS

The major aggregate projection results of the Timber Supply Model are summarized in tables 10-1 and 10-2. These results indicate that the TSM base case projects the total worldwide industrial wood harvest to increase to 1.7 billion m^3 in 2000 and 2.0 billion m^3 by the year 2035. Over the entire fifty-year period and the fifteen-year subperiod to 2000, the real growth of real wood prices is projected to grow at an average annual rate of 0.2 percent. The TSM high demand scenario generates somewhat higher growth rates, with the harvest in 2000 projected at 1.8 billion m^3 and that in 2035 projected at 2.3 billion m^3. The real price growth rate projected for the high demand scenario is 1.2 percent annually for the period between 1985 and 2000 and 1.3 percent annually for the entire fifty-year period to 2035.

Harvest Comparisons

To provide perspective it is useful to compare the projections of the TSM with other projections and forecasts of industrial wood demand and production. Table 10-1 also presents the projections of several forecasts made by authoritative groups over the past several years. For harvest projections to the year 2000, our TSM base-case projection of 1.7 billion m^3 is the lowest of those presented. However, it is only slightly lower than that of the FAO Industry Working Party (1.8 billion m^3), the IIASA Global Trade Model (1.8

Table 10-1. Forecasts of Industrial Roundwood Demand

Organization/study[a]	Year made	Projected to year:	Forecast volume (billions of m^3)	Implicit growth rate[b] (percent/year)
Food and Agriculture Organization	1982 (high)	2000	2.6	3.7
	(low)	2000	2.3	2.9
Food and Agriculture Organization, Industry Working Party	1979	2000	1.8	1.2
SRI	1979	2000	1.9	1.6
World Bank	1978	2000	2.8[c]	4.2
		2025	5.9[c]	3.4
International Institute for Applied Systems Analysis (IIASA) GTM	1987	2000	1.8	1.2
	1987	2030[d]	2.6	1.2
RFF TSM				
Base case	1988	2000	1.7	0.8
	1988	2035	2.0	0.6
High demand	1988	2000	1.8	1.2
	1988	2035	2.3	0.9

Source: Adapted from David W. K. Boulter, "Global Supply-Demand Outlook for Industrial Roundwood," *Proceedings of the National Forest Congress,* April 9 (Ottawa, Canadian Forestry Association, 1986) p. 5.

[a]SRI International and World Bank data from Boulter (1986). FAO 1982 data from FAO, "World Forest Products: Demand and Supply 1990 and 2000," FAO Forestry Paper no. 28 (Rome, FAO, 1982). FAO Industry Working Party data from FAO, "FAO World Outlook for Timber Supply," phase V (Rome, FAO, 1979). IIASA data from M. Kallio, D. P. Dykstra, and C. S. Binkley, eds., *The Global Forest Sector: An Analytical Perspective* (New York, Wiley, 1987).

[b]From a base of 1.5 billion m^3 in 1985.

[c]Based on constant real prices.

[d]The IIASA GTM projections to 2030 are not reported in *The Global Forest Sector.* They were presented, however, at numerous meetings and in working papers such as "The Global Forest Model" (unpublished draft by Luke Popovich, May 1985).

Table 10-2. Projected Real Price Rises for Industrial Wood

Study	Year made	Projected to year:	Real price growth rate (annual percentage)[a]
RFF TSM			
Base case	1988	2000	0.2
High demand	1988	2000	1.2
		2035	1.3
IIASA GTM			
Eastern United States			
Sawlogs, conifer		2000	5.9
Pulplogs, conifer		2000	3.5
Sawlogs, nonconifer		2000	4.7
Pulplogs, nonconifer		2000	0.0
Western Europe			
Sawlogs, conifer		2000	0.3
Pulplogs, conifer		2000	3.2
Sawlogs, nonconifer		2000	1.3
Pulplogs, nonconifer		2000	1.8

Source: IIASA data from D. P. Dykstra and M. Kallio, "Base Scenario," chap. 28 in M. Kallio, D. P. Dykstra, and C. S. Binkley, eds. *The Global Forest Sector: An Analytical Perspective* (New York, Wiley, 1987).

[a]From 1980 to 2000.

billion m^3), and the Stanford Research Institute (1.9 billion m^3). Our high-demand scenario of 1.8 billion m^3 is, of course, within this grouping. In fact, these four studies produce remarkably close projections for the year 2000. Our TSM projections are, however, substantially lower than those of the FAO in its 1982 study (2.3 billion to 2.6 billion m^3) and the World Bank study (2.8 billion m^3) for the year 2000.

The only projections presented that go well beyond the year 2000 are those of our TSM, the World Bank, and IIASA. Here the differences become large. The World Bank projection of demand of 5.9 billion m^3 for the year 2025 is more than twice as large as our TSM base-case projection of harvests of 2.0 billion m^3 for 2035 (or 2.3 billion m^3 for our high-demand scenario) and the IIASA projection of 2.6 billion m^3 for the year 2030. The implicit growth rate of the World Bank study is more than four times that of our TSM: 4.2 percent annually as compared with our 0.8 percent. Because our projections show only a very modest price rise (as discussed later), our projections are conceptually comparable to those of the World Bank, both being essentially based on fixed prices.

In summary, our TSM projections of overall timber production in the year 2000, though conservative, fall within a range of commonly accepted estimates. For the longer term that carries into the first three decades of the twenty-first century, our projections would surely be conservative when compared to other projections such as those of the World Bank and even of IIASA. However, as has been noted, the assumptions of the TSM have the effect of biasing the harvest projections (but not real price estimates) downward when applied over very long time periods.

Price Projections

Most of the global studies focus on harvest volumes, but our TSM and the IIASA GTM also project an intertemporal profile of real prices for timber. The results of the TSM base-case and high-demand scenarios, together with the IIASA base-case projections, are presented in table 10-2 for the period from 1980 to 2000. The TSM, which generates a price projection for a single composite harvest, shows an average real price rise of 0.2 percent for the base case and 1.2 percent for the high-demand scenario over the period from 1985 to 2000. The IIASA price projections, in contrast, are disaggregated by region and by log type. The real price growth rates reported in the table are those for the eastern United States and western Europe and are highly variable, with those of the eastern United States running from zero growth for nonconifer pulpwood to 5.9 percent for conifer sawlogs. For western Europe the growth rates run from 0.3 percent for conifer sawlogs to 3.2 percent for conifer pulpwood logs.

Implications of the Projections

Although considerable differences in the projections exist, they are not as large as they might appear at first. The extreme harvest projections are those of the FAO and World Bank, which were done in the late 1970s or very early 1980s. To a large extent the stagnation experienced in the early 1980s has swept away the perspectives on which the relatively rapid harvest expansions were based. As Boulter (1986, p. 5) has noted, by the mid-1980s views had been tempered by the experience of the first half of the decade, and the tendency was to accept a forecast in the range of 1.8 billion m^3 to 2.0 billion m^3.

Although price projections of industrial wood in the form of logs or stumpage are fairly common for regional studies (for example, see USDA, Forest Service, 1987), only two of the global studies cited here—the TSM and the IIASA study—undertake serious price projections. A direct comparison of the results is difficult because the TSM approach is aggregate and the IIASA approach is highly disaggregate. In a recent paper, however, Binkley and Vincent (1988) used the results of the TSM and the IIASA study as well as other regional studies to examine the real prices of sawtimber in the U.S. South. In their study they compared a real price growth rate for conifer sawtimber stumpage of 3.4 percent annually, derived from the IIASA study, with their real price estimate of conifer sawtimber stumpage growth of 0.5 percent annually for the TSM base-case and of 1.9 percent for the high-demand scenario.

If these are accepted as reasonable surrogates for the overall price trends inherent in the projections of the two studies, the real price estimates of the IIASA study grow at almost seven times the TSM rate for the base-case and twice as fast for the TSM high-demand scenario. An important finding is that although the TSM high-demand scenario and the IIASA study projections provide essentially equal harvest volumes in the year 2000, the real price increases of the IIASA model are almost twice those of the TSM. This comparison makes it clear that the TSM relies more on supply-side responses to price and less on the inhibiting effects of price on demand.

WHY THESE FINDINGS?

The major finding of this study—that under reasonable assumptions about future conditions, industrial wood real prices are unlikely to exhibit substantial long-term increases—flies in the face of the once-common view in forestry that envisages continuing economic scarcity reflected in sharply rising real prices for industrial wood. This view is inherent in the assumption sometimes used for investment analysis in forestry, that is, that real prices of industrial wood will rise into the indefinite future. Furthermore, this view contrasts sharply with the projections of the IIASA GTM, which projects a real price growth rate over the next two decades that is seven times as rapid as the TSM base case and almost twice as rapid as the TSM high-demand scenario. The findings of our TSM are, however, generally consistent with those of the ECE/FAO in *European Timber Trends and Prospects to the Year 2000 and Beyond* (ECE/FAO, 1986), which anticipates no serious upward movement in real prices in the foreseeable future.

The differences between this study's generally optimistic view of future industrial wood availability and the more pessimistic view represented by the conventional wisdom and some other studies can be traced generally to three factors. First, this study suggests that the most likely future scenario is modest increases in demand for industrial wood into the indefinite future. Second, the greater responsiveness of supply to real price increases in the TSM is the result of both responses to silvicultural investments and the availability of numerous "backdrop" regions that can be brought into production if real prices rise. This supply responsiveness generates a high-demand scenario where the harvest volumes are identical to those of the IIASA model while real price growth rates are only one-half as great. Finally, the introduction of the role of technological change into the analysis—both formal and informal—allows a more reasonable assessment of the degree to which demand for the wood input may increase over time and the extent to which supply can respond over the long term.

The modest growth in demand in the base case is due in part to a maturing of the industrial economies of the Northern Hemisphere but also, and importantly, to the moderating effect that wood-saving technological change is likely to have on demand for industrial wood. Wood-saving technology will allow the demand for the raw material to grow indefinitely less rapidly than the demand for final wood products. Also, as wood-using industrial economies reach maturity, their populations will tend to stabilize and their housing stocks will approach levels consistent with long-term requirements largely for replacement and repair. Hence demand for industrial wood, particularly solidwood, will grow only slowly.

The major source of any rather dramatic increases in demand for industrial wood resources would almost certainly be the Third World economies. The dramatic growth over the next two or three decades from this source is, however, highly problematic. Even should these countries experience relatively rapid economic growth, the massive foreign debt burdens that many Third World countries face could constrain the extent to which they import wood and wood-using products for decades to come. These countries ultimately may increase their consumption of wood products; however, the foreign debt constraint makes it likely that many of them will rely heavily on their ability to meet these needs from domestic resources. Such a response would offset increases in demand with corresponding increases in domestic supplies. Although this outcome would result in a higher level of aggregate demand and supply than is simulated in the TSM, the net effect on real price and economic scarcity is likely to be largely offsetting, hence modest.

The analysis of chapter 5 as translated into the construction of the model suggests that technological change can affect economic supply as well as demand, as in chapter 6. Technological change has resulted in the past (and can be expected in the future to continue to result) in profound changes on the supply side by increasing physical production from the forests in numerous ways. Continuing improvements in forest management and the introduction of progressively superior genetically improved trees will cause yield enhancement over time. In addition, technology allows the economic utilization of previously noneconomic "underutilized" species and tree residuals. Although both effects would result in an outward shift of the industrial wood supply curve, the latter effect has not been built into the formal model. The effect of this misspecification of the formal model is to bias the harvest and real price projections downward.

In summary, technology can be viewed as working on both the demand curve and the supply curve. On the demand side wood-saving technology reduces the effect of increases in demand for the final product upon demand for the raw material. On the supply side such technology serves to expand supply through both wood-extending and yield-enhancing effects.

LONG-TERM WORLD TIMBER AVAILABILITY: SOME CONCLUSIONS

As noted in this book's introduction, industrial wood has been identified as one of the few natural resources that has been growing progressively more scarce in an economic sense, that is, experiencing a long-term rising real price over the past several decades or more. As noted, not all industrial wood has experienced this real price rise. Rather, the rise has been confined largely to wood that goes into solidwood products, particularly lumber and sawlogs. Other types of industrial wood, such as pulpwood and wood used for producing composite solidwood products, have not experienced the rising long-term trend in their real price. Also, as demonstrated in chapter 3, there is evidence that the long-term real price rise moderated considerably after 1950.

The historical experience is consistent with the view first formalized for forestry in the theoretical model developed by Lyon (1981) and refined and modified into our TSM. This view sees the exploitation of forests for industrial wood as a progression through the "foraging" stage of harvesting natural, old growth forests; rising real prices are associated with the drawing down of old growth stock. As the process continues the real price rise creates incentives first for forest management and ultimately for forest farming.

This view does not require a world exclusively in plantation forests to complete the process. Rather, varying site classes and locational considerations allow for all three forms of forestry—unmanaged, managed natural, and plantation forests—to be practiced simultaneously in a worldwide system. In addition, many old growth forests would remain unexploited because of accessibility costs and constraints. In the context of this conceptual system, the rise in real prices of wood will gradually diminish and level off. The introduction of technological change into this system will facilitate the process of price leveling and conceivably could generate falling long-term real prices such as have been experienced with agricultural commodities and other natural resources.

It is clear that even in view of the acceptance of the conceptual view embodied in the TSM, the historical evidence does not yet allow for a definitive assertion that long-term real price rises for industrial wood are a thing of the past. However, numerous bits of evidence suggest that the old growth drawdown process is well advanced and that the world economy is in transition toward a much greater reliance on forest management and timber growing. For example, the world's timber supply is increasingly being met from second- and third-generation forests. In the United States, the center of gravity of the timber-producing industrial forest is shifting rather dramatically from the old growth Pacific Northwest region to the forest regrowth and plantation forests of the South. In addition, intensively managed plantation forests are being planted, managed, and harvested on favorable sites in numerous locations around the globe.

Although governments support much of this activity, the higher real prices experienced for some types of industrial wood and improvements in the technology of tree growing both contribute to this phenomenon. Apart from the newly emerging producers of Brazil, Chile, New Zealand, South Africa, Spain, and Portugal, plantations are being established at very rapid rates in places such as the U.S. South. Almost all the plantation activity is of recent vintage; most of it has taken place after 1955. Hence we are now actually observing a structural shift away from old growth and toward plantations in a manner consistent with the conceptual model.

Have we reached the peak level of real prices for industrial wood and largely completed the transition? Probably not. Some types of higher-quality sawlogs may well experience some future increases in real price over the next couple of decades. However, the rate of price increase has moderated in recent decades, and the continuation of further moderation seems reasonable to expect. Surely there is little reason to expect pulpwood real prices to increase systematically in the future when they did not do so during the past eighty years. Furthermore, there is little reason to expect any long-term price rises in the low-quality industrial wood used in the solidwood composites. Rather, the composites are likely to put downward pressure on the prices of solidwood products requiring sawlogs, hence downward pressure on the real prices of sawlogs themselves, thereby dampening any real price increases that might occur. In short, for the reasons just listed, the future long-term rate of growth of the real prices of wood may be anticipated to be less rapid than in the past.

SUMMARY OF MAJOR FINDINGS

The major finding of this study is that under a reasonable set of assumptions regarding demand and supply, the market demand for industrial wood over the next two to three decades, and perhaps much longer, can be accommodated adequately by likely forthcoming economic supplies without large increases in the real price of industrial wood. Although this real price rises more rapidly with the TSM projections of the high-demand scenario, these price rises are still relatively modest.

Historical data suggest that the rate of increase in prices and harvest volumes has declined in the post-1950 period. This finding is consistent with a situation in which the global system is nearing the completion of a transition to a more steady-state situation where real timber prices would be expected to exhibit less growth.

The structure of the TSM suggests that if there is a high degree of interrelatedness among regional markets, "shocks" on the system originating in one region, such as a large tax increase in the United States or a dramatic increase in Soviet timber harvests, are largely offset in the long run by accommodating adjustments in production, investment, and harvests in other regions.

A high degree of regional interrelatedness has both advantages and disadvantages. Interrelatedness allows the system as a whole to mitigate much of the aggregate effect of a large local shock, but this same interrelatedness is likely to exacerbate the effects of such shocks on an individual region. This is because the price rise associated with an output-reducing shock in a closed regional system and advantageous to local producers is dramatically dampened in an open system by the responsiveness of the other, now advantaged, regions.

The model suggests that the forester's concept of an even-aged, fully regulated forest needs some revision when applied with an economic criterion to a global system in which various regions have differing initial forest inventories, differing site class productivities, and differing locational advantages. Such a system is likely to generate a series of very different regional forests, few if any of which are individually even aged or regulated. Collectively, however, the global whole produces a smooth flow of timber, with each regional forest contributing (albeit quite differently) to that synchronized whole.

The model allows for a variety of management regimes and intensities within the global system. Well-situated, high-productivity forestlands are likely to experience intensive forestry while poorly located and/or low-productivity sites will experience little management. Old growth, unmanaged forests will persist undisturbed in inaccessible regions.

The model demonstrates that over the long term, timber regions are likely to wax and wane as producing regions, and there will probably be a shift in production away from the old growth regions and toward the regions of high growth. However, all seven regions examined in this study will continue to be significant timber producers into the indefinite future.

A CAVEAT

The foregoing analysis represents an attempt to grapple with an essentially unknowable question—that of timber supply in the future. As we have tried to make clear throughout, our results are not and cannot be definitive.

Numerous surprises could intervene to disrupt the timber market in ways we cannot anticipate. As we noted in the introduction, certain environmental changes could negate the assumptions and underlying production functions of this study and so invalidate the projections and analysis.

The effect that environmental changes are likely to have on our projections relates to the size, timing, and focus of the environmental disturbance on the forests that are relevant to future economic timber supply. For example, small localized environmental disturbances are unlikely to affect significantly our global projections. Similarly, larger environmental changes that affect forests that are not particularly important sources of economic supply are likely to have only minimal effects on the projections. In contrast, environmental changes that focus large effects on regions that *are* important sources of economic supply could alter significantly the system that has been modeled, rendering the assumptions inappropriate and undermining the validity of the analysis and projections.

Given these considerations we would not expect that environmental disruptions generated by current levels of tropical deforestation or by moderate but localized forest dieback resulting from air pollution are likely to disrupt seriously the assumptions of the TSM or our aggregate analysis and projections. Environmental disruptions could, however, significantly affect the composition of that output if, for example, one region were severely disrupted. Also, preservationist policies that withdraw large areas of timber from harvest could disrupt the projections—again, probably more the composition than the aggregate. But if the current aggregate projections are badly in error, the cause is likely to be factors other than such environmental disruption. However, should the world experience a large, pervasive, pollution-induced forest dieback or a major global warming, the accuracy of the analysis and projections is likely to be affected. Such a major environmental change could so alter the dynamics of the forest system that some of the parameters and assumptions embodied in the production functions of the TSM model would no longer be relevant.

One important issue in determining whether these considerations affect the current TSM projections is timing. For example, to the extent that global warming would occur slowly or lies in the distant future, the assumptions of the analysis and projections retain their relevance; to the extent that warming would occur rapidly and soon, the effect of climate change on the projections and analysis is likely to be large.

Even given these uncertainties, however, we believe that the analysis, forecast, and simulations provide a useful starting point for an assessment of the future prospects for a particular natural resource—timber—and its market. We hope that this work will stimulate and perhaps even provoke its readers in order to encourage continuing attempts to understand better the issue of resource scarcity in general and of timber scarcity in particular.

REFERENCES

Binkley, Clark S., and Jeffrey Vincent. 1988. "Timber Prices in the U.S. South," *Southern Journal of Applied Forestry* vol. 12, no. 1 (February) pp. 12–18.

Boulter, David W. K. 1986. "Global Supply-Demand Outlook for Industrial Roundwood," *Proceedings of the National Forest Congress*, April 9 (Ottawa, Canadian Forestry Association).

Dykstra, D. P., and M. Kallio. 1987. "Base Scenario," pp. 629–647 in M. Kallio, D. P. Dykstra, and C. S. Binkley, eds., *The Global Forest Sector: An Analytical Perspective* (New York, Wiley).

Economic Commission for Europe/Food and Agriculture Organization of the United Nations (ECE/FAO). 1986. *European Timber Trends and Prospects to the Year 2000 and Beyond* (Geneva, ECE/FAO).

Food and Agriculture Organization of the United Nations (FAO). 1979. "FAO World Outlook for Timber Supply," phase V, prepared by a Joint Forestry and Industry Working Party for the Forestry Department (Rome, FAO).

Food and Agriculture Organization of the United Nations (FAO). 1982. "World Forest Products: Demand and Supply, 1990 and 2000," Forestry Paper no. 29 (Rome, FAO) p. 336.

Kallio, M., Dennis P. Dykstra, and Clark S. Binkley, eds., 1987. *The Global Forest Sector: An Analytical Perspective* (New York, Wiley).

Lyon, K. S. 1981. "Mining of the Forest and the Time Path of the Price of Timber," *Journal of Environmental Economics and Management* vol. 8, pp. 330–344.

SRI International. 1979. "The Outlook for the World's Forest Products Industries" (Palo Alto, SRI International).

U.S. Department of Agriculture (USDA), Forest Service. 1987. "The South's Fourth Forest" (Washington, D.C., USDA).

World Bank. 1979. EIS Paper no. 98, prepared by James Gammie (London, IBRD).

Appendixes

Introductory Note

The Timber Supply Model (TSM) incorporates physical and biological elements to provide a biological production function for each of the twenty-two land classes (appendix A), or basic units of production, in seven major timber-producing regions of the world.

Appendixes B through H present the data used to develop these production functions. Included is information on land class quality, location, accessibility, and area; growth yield functions by dominant species and land class; existing inventories and their age distribution; and suitability of the timber for sawlogs or pulping. Appendixes I through O present additional technical information incorporated in the model or about the model itself. The appendixes are organized as follows:

A. Disaggregation of seven major timber-producing regions into twenty-two land classes

B. Emerging Region forest data (land class 1)
C. U.S. Pacific Northwest forest data (land classes 2 through 5)
D. British Columbia forest data (land classes 6 and 7)
E. Nordic Europe forest data (land classes 8 and 9)
F. U.S. South forest data (land classes 10 through 17)
G. Eastern Canada forest data (land classes 18 through 21)
H. Asia-Pacific forest data (land class 22)

I. Cost information—on harvesting, domestic transportation, and international transportation—incorporated in the model

J. Exchange rate assumptions

K. The economy-wide interest rate, reflecting the opportunity cost of capital, introduced in the model

L. Discussion of the introduction of technology into the model

M. Description of the yield function, which plays an integral role in the model

N. Explanation of regeneration input, which is one of the two components of the yield function (the other component is age of the stand)

O. Presentation of detailed technical information on the Timber Supply Model

Appendix A
Regions and Land Classes

Regions/characteristics	Land class
Emerging region	1
U.S. Pacific Northwest, west of Cascades, normal access	2
U.S. Pacific Northwest, west of Cascades, limited access	3
U.S. Pacific Northwest, east of Cascades, normal access	4
U.S. Pacific Northwest, east of Cascades, limited access	5
Western Canada (British Columbia), normal access	6
Western Canada (British Columbia), limited access	7
Nordic Europe (southern)	8
Nordic Europe (northern)	9
U.S. South, site class I, normal access	10
U.S. South, site class I, limited access	11
U.S. South, site class II, normal access	12
U.S. South, site class II, limited access	13
U.S. South, site class III, normal access	14
U.S. South, site class III, limited access	15
U.S. South, site class IV, normal access	16
U.S. South, site class IV, limited access	17
Eastern Canada, Lake, pulp	18
Eastern Canada, Lake, sawlog	19
Eastern Canada, Boreal/Acadia, pulp	20
Eastern Canada, Boreal/Acadia, sawlog	21
Asia-Pacific	22

Appendix B
Emerging Region Forest Data

The emerging region (land class 1) contains a composite of the characteristics of newly established industrial forest plantations in several countries that use exotic species. These plantations have been established in countries of the Southern Hemisphere, the tropics, and the Iberian peninsula of Europe; these countries include Brazil, Chile, Venezuela, Australia, New Zealand, South Africa, Spain, and Portugal. There is one composite land class for the entire region. The emerging region has 6 million ha in its initial period and a pulpwood regime.

Our study of the emerging region begins with an initial endowment of forest inventories that reflects the situation in about 1985. Subsequently, various scenarios examine the implications of alternative rates of establishment of new industrial forest plantations.

The emerging region consists of several regions in which intensive plantation forestry is practiced. Rotations vary from seven to thirty years, and average annual growth rates run from 15 m^3/ha to above 30 m^3/ha. A composite yield function was constructed that captures "average" composite characteristics for the set of heterogeneous regions. These include a mean annual increment at harvest age of 20 m^3/ha and an economic rotation of fifteen years with a real interest rate of 6 percent. This yield function was constructed utilizing data from Roger A. Sedjo, *The Comparative Economics of Plantation Forestry* (Washington, D.C., Resources for the Future, 1983). See appendix M in this volume.

The initial inventory and age distribution were estimated using data from various sources as to the level and rate of establishment of new industrial forest plantations in the countries of the emerging region in the recent past and reductions to these from harvests and other sources. See Stephen E. McGaughey and Hans M. Gregersen, eds., *Forest-Based Development in Latin America* (Washington, D.C., Inter-American Development Bank, 1983); Roger A. Sedjo, "Plantations in Brazil and Their Possible Effects on World Pulp Markets," *Journal of Forestry* (November 1980); and Roger A. Sedjo, *The Comparative Economics of Plantation Forestry* (Washington, D.C., Resources for the Future, 1983).

The following table presents inventory data for the emerging region, land class 1.

Age of forest (in years)	Hectares (millions)	Age of forest (in years)	Hectares (millions)
1	0.54	9	0.38
2	0.52	10	0.36
3	0.50	11	0.34
4	0.48	12	0.32
5	0.46	13	0.30
6	0.44	14	0.28
7	0.42	15	0.26
8	0.40		

Appendix C
U.S. Pacific Northwest Forest Data

The Pacific Northwest has a total of four land classes east and west of the Cascades. The western area (land classes 2 and 3) contains 7.41 million ha; of these, 4.91 million ha in land class 2 have sawlog output, and 2.5 million ha in land class 3 also have sawlog output.

The eastern area contains 8.53 million ha; of these, 3.38 million ha in land class 4 have sawlog output, and 5.15 million ha in land class 5 also have sawlog output.

These data include both roaded and nonroaded areas but exclude wilderness and proposed wilderness areas. Because yield functions (see appendix M) were given in total rather than merchantable volumes, they were converted to merchantable volumes for the purposes of this study; the converted data appear below.

The land classes in the west of the region contain Douglas fir; the yield function is an average of site classes I through V weighted by relative land areas. Land classes 2 and 3 have the same yield function. See Richard E. McArdle, Walter H. Meyers, and Donald Bruce, "The Yield of Douglas Fir in the Pacific Northwest," *USDA Technical Bulletin* no. 201 (Washington, D.C., USDA, Forest Service, 1930) p. 23, and also Robert O. Curtis, Gary W. Clendenen, Donald L. Reukema, and Donald J. DeMurs, "Yield Tables for Managed Stands of Coast Douglas Fir," USDA Forest Service General Technical Report PNW-135 (Washington, D.C., USDA, 1982). These data were converted to merchantable equivalents using the information provided by Walter Meyers in his section on the application of yield tables in *USDA Technical Bulletin* no. 201.

The land classes in the east of the region contain Ponderosa pine; the yield function is an average of site classes I through V weighted by the relative land areas. Land classes 4 and 5 use an identical yield function. See Walter H. Meyers, "Yield of Even-Aged Stands of Ponderosa Pine," *USDA Technical Bulletin* no. 630 (Washington, D.C., USDA, 1938). A program for estimating merchantable yields from total yield was developed by Katherine Tunis.

Forestry inventory data for the Pacific Northwest were obtained from the USDA Forest Service computer tapes (Tape RFF356 from letter of Katherine Tunis, Resources for the Future, to Kenneth Lyon, July 15, 1982).

The following table presents age group inventory data for Pacific Northwest land classes 2 through 5. Data for land classes 2 and 3 are in the column labeled West, and data for land classes 4 and 5 are in the column labeled East. Ages 1 through 80 are the normal-access hectares, land classes 2 and 4; the remaining hectares are in the limited-access land classes 3 and 5.

Age of forest (in years)	Hectares (millions)	
	West	East
1–10	1.0897	0.6794
11–20	0.7950	0.4890
21–30	0.6037	0.1219
31–40	0.5988	0.1028
41–50	0.6302	0.3055
51–60	0.5475	0.4659
61–70	0.3626	0.5807
71–80	0.2877	0.6352
81–90	0.2699	0.7188
91–100	0.1977	0.5958
101–110	0.1590	0.5071
111–120	0.1118	0.4234
121–130	0.1149	0.4152
131–140	0.0743	0.3626
141–150	0.0716	0.3273
151–160	0.0927	0.2323
Over 160	1.4144	1.5672

Appendix D
British Columbia Forest Data

British Columbia has two land classes: a normal-access and a limited-access class. Its total area comprises 32 million ha. The normal-access land class (6) has 17.7 million ha. The limited-access land class (7) has 14.3 million ha.

The yield function is the same for both land classes. The distinction between these land classes derives from differing access costs, which result in differing harvest costs. All the land considered is characterized as "operable" by official Canadian sources. (For the yield function see appendix M.) The Ponderosa pine on site 4.5 is 50 percent stocked. See Walter Meyers, "Yield of Even-Aged Stands of Ponderosa Pine," *USDA Technical Bulletin* no. 630 (Washington, D.C., USDA, 1938), and G. M. Bonner, *Forest Resource Inventory 1981* (Department of the Environment, Canadian Forestry Service, Forest Statistics and Systems Branch, 1982).

An estimate of the initial existing inventory for this region was derived from the land area, the average site class (4.5), the initial age distribution (see Bonner), and the yield function. The initial age distribution was 4.5 percent of land under 30 years old, 39.5 percent of land between 30 and 70 years, and 56 percent of land over 70 and up to 120 years.

The following table presents age group data for British Columbia, land classes 6 and 7. Ages 1 through 80 are the normal-access hectares, land class 6; the remaining hectares are in the limited-access land class, 7.

Age of forest (in years)	Hectares (millions)
1–10	0.4847
11–20	0.4847
21–30	0.4847
31–40	3.1643
41–50	3.1643
51–60	3.1643
61–70	3.1643
71–80	3.5778
81–90	3.5778
91–100	3.5778
101–110	3.5778
111–120	3.5778

Appendix E
Nordic Europe Forest Data

The Nordic countries of Sweden, Finland, and Norway have two land classes: the north and south Nordic classes. (The Nordic region was divided into north and south classes to reflect the different access and climate conditions associated with these different locations.) The entire region comprises 50.43 million ha. The south Nordic class (8) has 22.29 million ha with sawlog output; the north Nordic class (9) has 28.14 million ha with pulpwood output.

The yield functions (see appendix M) for Norway spruce were developed for average site conditions for both the north and south Nordic regions using Swedish data (appendix M). See Nils-Erik Nilsson, "An Alley Model for Forest Resources Planning" (Jonkoping, Sweden, The National Board of Forestry, n.d.), and *Sveriges Officiella Statistik, 1981* (Skogsstyrelsen, Jonkoping, Skogsstatistik arsbok, 1982).

The estimates of initial inventory volumes and age distribution were obtained from Swedish and Finnish data and extrapolated to the rest of the Nordic region. See *Sveriges Officiella Statistik, 1981* and *Yearbook of Forest Statistics: 1975* (Helsinki, Official Statistics of Finland, SVII A, 1977).

The following table presents data for Nordic Europe (land classes 8 and 9).

	Hectares (millions)	
Age group	South	North
1–10	3.3660	3.3000
11–20	2.2440	2.3320
21–30	1.3420	1.2320
31–40	1.3420	1.2320
41–50	1.9800	1.2320
51–60	1.9800	1.2320
61–70	2.2000	1.4080
71–80	2.2000	1.4080
81–90	1.6500	2.2660
91–100	1.6500	2.2660
101–110	0.8360	2.4860
111–120	0.8360	2.4860
121–130	0.3300	1.3860
131–140	0.3300	1.3860
141–150	0	0.6380
151–160	0	0.6380
Over 160	0	1.2100

Appendix F
U.S. South Forest Data

The U.S. South contains site classes I through IV. There are two land classes per site. Site I contains 20.65 million ha. Land class 10 (normal access) has 16.14 million ha with pulpwood output; land class 11 (limited access) has 4.51 million ha with sawlog output.

Site II has 20.18 million ha. Land class 12 (normal access) has 15.97 million ha with sawlog output; land class 13 (limited access) has 4.21 million ha with pulpwood output.

Site III has 19.82 million ha. Land class 14 (normal access) has 9.53 million ha with pulpwood output; land class 15 (limited access) has 1.29 million ha with sawlog output.

Finally, Site IV has 16.27 million ha. Land class 16 (normal access) has 9.46 million ha with pulpwood output, and land class 17 (limited access) has 6.81 million ha with sawlog output.

Site I consists of Loblolly pine. The same yield function, site index 90 natural, is used for both normal-access and limited-access land classes. (For the yield functions see appendix M.) See Harold F. Burkhart, Robert C. Parker, and Richard G. Oderwald, "Yields for Natural Stands of Loblolly Pine," Pub. no. FWS-2-72 (Blacksburg, Virginia Polytechnic Institute and State University, Division of Forestry and Wildlife Resources, December 1972).

Site II also contains loblolly pine, and the same yield function, site index 80 natural, is used for both normal- and limited-access land classes. (See Burkhart and coauthors.) Site III too contains loblolly pine and uses the same yield function, site index 60 natural stand, for both classes. (See Burkhart and coauthors.)

Finally, Site IV contains mixed hardwood. A relatively slow-growing yield function is used to capture the features of this site class for both normal- and limited-access land classes. (See appendix M.)

The forest inventory data for the U.S. South were provided by the Forest Service's Southern and Southeastern Experiment Stations. The analysis assumes that if regions I, II, and III were harvested, they would be regenerated in pine or some species with growth characteristics similar to those captured in the pine yield function used.

The following table presents age group data for the U.S. South, land classes 10 through 17. Data for land classes 10 and 11, 12 and 13, 14 and 15, and 16 and 17 are in the columns labeled Site I, Site II, Site III, and Site IV, respectively. Ages 1 through 50 are the normal-access hectares, land classes 10, 12, 14 and 16; the remaining hectares are in the limited-access land classes, 11, 13, 15, and 17.

Age of forest (in years)	Hectares (millions)			
	Site I	Site II	Site III	Site IV
1–10	3.3933	2.6931	2.0877	1.4679
11–20	2.5514	2.1968	2.1180	1.3398
21–30	3.2191	3.3179	2.2296	1.6130
31–40	3.7403	4.3286	1.8428	2.2577
41–50	3.2352	3.4330	1.2519	2.7771
51–60	2.8497	2.3727	0.6949	2.3927
61–70	0.9415	0.9936	0.3329	1.8810
71–80	0.3855	0.4151	0.1427	0.9966
81–90	0.3356	0.4319	0.1152	1.5363

Appendix G
Eastern Canada Forest Data

Eastern Canada has two areas: the Boreal/Acadia (B/A) and Lake areas. Each has two land classes. Those of the B/A area have 57.3 million ha; 30.37 million ha (land class 20) are in pulpwood and 26.93 million ha (land class 21) are in sawlog. The land classes of the Lake area have 27.79 million ha; 12.6 million ha (land class 18) are in sawlog and 11.18 million ha (land class 19) are in pulpwood. All this land lies in operability class 1, which is the most easily accessible. It includes slightly over one-half of eastern Canada's total forestland. (For yield functions see appendix M.)

Land classes 20 and 21 of Site II of the B/A area contain balsam fir. See Michael Boudoux, "Tables de rendement empiriques, pour l'epinette noire, le sapin baumier, et le pin gris au Quebec" ("Empirical Yield Tables for Black Spruce, Balsam Fir, and Jack Pine in Quebec") (Canadian Forestry Service, Center for Forest Research of Laurentides, 1978).

Land classes 18 and 19 of Site II of the Lake area contain red pine. See W. L. Planski, "Normal Yield Tables for Black Spruce, Jack Pine, Aspen, White Birch, Tolerant Hardwoods, White Pine, and Red Pine for Ontario" (Ontario Department of Lands and Forests, June 1960, repr. January 1971).

The estimate of initial existing forest inventory was derived from the land area, average site class (II), estimated initial age distribution, and yield function. The initial age distributions were 10 percent of the land area in age class 0 to 10 years; 43 percent of the land area in age class 10 to 50 years; and 47 percent of the land area in age class 50 and above. The existing land area was spread evenly over the appropriate age intervals. The yield function used was that which would give volumes equal to one-half of a fully stocked stand. These estimates were checked against average volumes by age for the region. See G. M. Bonner, "Canada's Forest Inventory 1981" (Department of the Environment, Canadian Forestry Service, Forestry Statistics and Systems Branch, 1982).

The following table presents age group data for Eastern Canada, land classes 18 through 21. Data for land classes 18 and 19 are in the column labeled Lake, and data for land classes 20 and 21 are in the column labeled Boreal/Acadia. Ages 1 through 50 are the normal-access hectares, land classes 18 and 20; the remaining hectares are in the limited-access land classes, 19 and 21.

Age of forest (in years)	Hectares (millions)	
	Lake	Boreal/Acadia[a]
1–10	2.3790	5.7300
11–20	2.5570	6.1600
21–30	2.5570	6.1600
31–40	2.5570	6.1600
41–50	2.5570	6.1600
51–60	0.9320	2.2440
61–70	0.9320	2.2440
71–80	0.9320	22.4425
81–90	0.9320	0
91–100	0.9320	0
101–110	0.9320	0
111–120	0.9320	0
121–130	0.9320	0
131–140	0.9320	0
141–150	0.9320	0
151–160	0.9320	0
Over 160	0.9320	0

[a]Beyond age 71 the Boreal/Acadia species achieve little additional volume. Therefore, all age groups above 71 are combined into age group 71–80.

Appendix H
Asia-Pacific Forest Data

The Asia-Pacific region is an aggregate of the tropical hardwood forests of Indonesia (Sumatra and Kalimantan), Malaysia (East and West), and the Philippines, all combined into one land class (22) comprising 58.31 million ha.

The harvest regime in the Asia-Pacific region entails selective harvesting of large, old growth, tropical hardwoods on a forty-year cutting cycle. This is a conservative estimate of the usual anticipated cycle. The younger, smaller trees (less than 60 cm in diameter) are left to continue their growth and are harvested at some later time. This process is designed to continue indefinitely. To approximate this behavior a yield function with a forty-year rotation (at the 6 percent operating rate of interest) was constructed. (See appendix M for the yield function.) The mean annual increment is 1 m^3/ha. See C. E. M. Keil, "Logging and Log Processing in Indonesia," a forest-sector input study for a World Bank working document (Washington, D.C., IBRD, 1978); M. S. Ross, "How Much Production Forestland Should Indonesia Plan to Keep?" paper presented at the First ASEAN Forestry Congress (Manila, Philippines, ASEAN, October 1983); and Food and Agriculture Organization of the United Nations (FAO), "Tropical Resources Assessment Project: Forest Resources in Tropical Asia," UN 32/6, 1301–78–04, Technical Report no. 3 (Rome, FAO, 1981).

The age distribution and volumes of the initial forest inventory were those of Bambang P. Adiwiyoto, "A Discrete Time Optimal Control Model for Optimization in Plywood Industry: A Case Study of Southeast Asia," Ph.D. dissertation (Logan, Utah, Utah State University, 1984).

The following table presents inventory data for the Asia-Pacific region, land class 22.

Age (years)	Hectares (millions)	Age (years)	Hectares (millions)	Age (years)	Hectares (millions)
1	2.19	14	1.25	27	0.30
2	2.53	15	1.11	28	0.25
3	2.52	16	0.85	29	0.21
4	2.67	17	0.75	30	0.21
5	2.57	18	0.73	31	0.19
6	2.36	19	0.66	32	0.19
7	1.92	20	0.57	33	0.20
8	2.28	21	0.51	34	0.80
9	2.38	22	0.49	35	0.70
10	1.96	23	0.43	36	0.80
11	1.67	24	0.38	37	0.40
12	1.61	25	0.35	over 38	19.98
13	1.44	26	0.33		

Appendix I
Harvesting, Domestic Transportation, and International Transportation Costs

Wood extraction and transport costs can be divided by (1) the cost of harvest, (2) the cost of transportation to mill, (3) forest access costs, and (for the purposes of this model), (4) international transportation costs. In addition, in an international model the effect of changes in the exchange rate structure on the dollar cost of domestic expenditure must also be considered.

HARVESTING AND DOMESTIC TRANSPORTATION

Table I-1 presents marginal cost estimates in dollars for each of the four cost components just mentioned for each of the model's twenty-two land classes. These dollar costs reflect the intermediate exchange rate structure that is used in the base scenario (see appendix J).

The first column lists land class numbers, and the second presents harvesting costs for each. These costs were developed by adapting the regional data provided in Roger A. Sedjo, *The Comparative Economics of Plantation Forestry* (Washington, D.C., Resources for the Future, 1983), for the intervening inflation as well as specific features of the land class such as terrain, log size, and other factors that affect harvesting costs.

The third column presents transportation costs to the mill, again adapted for changes in the price level and local factors. The fourth column provides estimates of road access costs developed from such costs for the U.S. western forests together with regionally specific topographic and other information. See Michael D. Bowes, John V. Krutilla, and Paul B. Sherman, "Forest Management for Increased Timber and Water Yields," *Water Resources Research* vol. 20, no. 6 (June) pp. 655–663.

INTERNATIONAL TRANSPORTATION COSTS

The inclusion of international transportation costs as the fifth column of table I-1 requires some explanation. The model is designed to examine potential global timber supply, hence some mechanism is needed to ensure that the costs of transportation beyond the mill to the global market are appropriately considered. To accomplish this we have assumed that all final wood products are marketed in one of three regional markets: the eastern United States, western Europe, or Japan/East Asia. Although the final product is some form of processed product, the international transportation cost can be built into the model by adding a factor for the equivalent cost of the international transportation of the unprocessed wood embodied in the final product.

For example, on the basis of the most common current technology, 1 ton of wood pulp embodies 4.7 m^3 of pulpwood. If the cost of transporting 1 ton of wood pulp from a mill to the nearest regional market is $100, the international transport cost is $100/4.7 or $21.28.

For each of the twenty-two land classes an estimate was made of the international transport costs of a cubic meter of embodied wood to the lowest-cost region. Wood pulp was used as representative of final wood products. These estimates were developed from data presented in Sedjo, *Comparative Economics*, p. 141. An important implication of this formulation is that the timber harvest price projected by the model is the mill price plus the cost of transporting a cubic meter of wood embodied in wood pulp. For our purposes this is a proxy of the price of timber as transacted in a unified world market.

Table I-1. Harvesting, Domestic Transportation, and International Transportation Costs

Land class [1]	Harvest costs (dollars per hectare) [2]	Domestic transport costs (dollars per hectare) [3]	Access costs (dollars per hectare) [4]	International transport costs (dollars per hectare) [5]
1	6.00	4.50	2.63	8.50
2	10.00	7.00	4.00	7.30
3	11.00	9.00	5.00	7.30
4	11.00	7.00	5.00	8.50
5	12.00	7.00	8.00	8.50
6	9.50	6.65	5.70	7.30
7	19.00	12.35	7.60	7.30
8	7.20	4.80	2.80	4.50
9	12.80	8.80	3.20	9.00
10	8.00	6.00	3.50	5.00
11	11.00	7.00	4.00	5.00
12	9.00	7.00	3.50	5.00
13	12.00	10.00	5.50	5.00
14	13.00	10.00	4.00	5.00
15	15.00	12.00	5.00	5.00
16	14.00	12.00	5.00	5.00
17	16.00	14.00	6.00	5.00
18	11.40	8.55	3.80	5.00
19	13.30	13.30	7.60	5.00
20	11.40	10.45	4.75	5.00
21	13.30	17.10	8.55	5.00
22	10.50	6.75	4.50	10.00

Appendix J
Exchange Rates

The TSM is not designed to project international trade flows, but it does deal with many countries with monies that are differently denominated. The need for a common unit of account requires that the model denominate all costs and prices in a common currency. The U.S. dollar has been chosen for this purpose, and all costs and prices in other currencies have been converted into U.S. dollars.

The question arises as to which exchange rate to use, as the exchange rate (or, more accurately, the structure of exchange rates) is changing over time. In this study the exchange rate structure used in the construction of the base case reflects the intermediate exchange rate structure. However, it is not certain or even likely that the dollar will maintain this position in the future. Because the exchange rate can have a major effect on the structure of future timber harvests and investments, this study examines the implications for the model's projections of alternative exchange rate structures.

The three alternatives are presented in table J-1. The first column is land classes. The second column reflects a weak dollar situation similar to that of 1980. The third column, which is simply a vector of 1s, reflects the intermediate dollar position used in the development of the base-case scenario; it is approximately reflective of the early 1986 structure. The fourth column is the strong dollar scenario roughly reflecting an exchange rate structure similar to that experienced in early 1985.

Table J-1. Exchange Rate Structures

Region	Land class [1]	Weak dollar [2]	Intermediate dollar [3]	Strong dollar [4]
Emerging	1	1.33	1.00	0.67
U.S. Pacific	2	1.00	1.00	1.00
Northwest	3	1.00	1.00	1.00
	4	1.00	1.00	1.00
	5	1.00	1.00	1.00
British	6	1.05	1.00	0.95
Columbia	7	1.05	1.00	0.95
Nordic	8	1.25	1.00	0.75
Europe	9	1.25	1.00	0.75
U.S. South	10	1.00	1.00	1.00
	11	1.00	1.00	1.00
	12	1.00	1.00	1.00
	13	1.00	1.00	1.00
	14	1.00	1.00	1.00
	15	1.00	1.00	1.00
	16	1.00	1.00	1.00
	17	1.00	1.00	1.00
Eastern	18	1.05	1.00	0.95
Canada	19	1.05	1.00	0.95
	20	1.05	1.00	0.95
	21	1.05	1.00	0.95
Asia-Pacific	22	1.33	1.00	0.67

Appendix K
The Interest Rate and the Sawlog-Pulpwood Mix

Most economic assessments of future costs and returns depend heavily on the choice of interest (discount) rate. Although this study is not an exception, we should also note that the basic long-term projections do not change dramatically for nominal changes in the choice of interest rate. The interest rate used for the study was a real 6 percent. This rate helped determine not only the rotation period of the pulpwood harvests but also the amount of investment to be made in regeneration for both sawtimber and pulpwood regimes. However, some modifications were required to capture the real-world differences in pulpwood and sawlog rotations.

The major outputs of industrial forests traditionally have been sawlogs and pulpwood. This is gradually changing with the advent of new technologies and new products; however, the sawlog-pulpwood dichotomy is still a useful characterization. It is well known that the usual Faustmann determination of rotation ignores the reality that sawlogs, which tend to be larger logs, command a greater value per unit of wood in the market. For this study, which treats industrial wood in the aggregate, some mechanism is needed to differentiate between the two types of outputs to allow for different rotation lengths.

The model utilizes a Faustmann-type mechanism to solve for the rotation length. In the absence of some type of adjustment, the solution procedure would generate rotation lengths for every region that would be of the short pulpwood type, as a single price for an undifferentiated final product was used. To overcome this difficulty and generate a realistic mix of sawtimber and pulpwood rotation lengths (hence harvest timings), the twenty-two land classes have been designated as being primarily either sawtimber or pulpwood. The designations are based on an assessment of the land classes' experience and potential. For regions designated as pulpwood, the model is allowed to solve for the rotation utilizing the economywide real interest rate of 6 percent. For the regions designated as primarily for sawtimber production, the real interest rate used for the solution of the rotation is reduced to 1.5 percent. The effect of this modification is to lengthen the sawtimber rotation to provide a solution that approximates the longer real-world sawtimber rotations.

Appendix L Introduction of Technology into the Model

Technology is introduced into the model in the form of wood-saving technological change through the demand side, as discussed in this book, and through the supply side through an outward shift in the yield function to reflect more rapid tree growth. In introducing technological change into the yield function, we envisioned it most easily as a genetic improvement that enhances the growth properties of the seedling. We have assumed that technology in tree growing (biotechnology) is progressing initially at a rate of 0.5 percent annually. Technology is embodied in the yield function and the current vintage introduced into the forest at the time of regeneration. Thus the vintage of the technology embodied in a forest is that which existed at the time of regeneration. Because technology is embodied in regeneration, technological improvement will not take place in the absence of positive expenditures for regeneration. Thus a completely naturally regenerated forest will not embody technological change.

For the full effect of technological change to be embodied in the forest, $500 of investment in regeneration is required. Between zero and $500, the 0.5 percent growth is proportionally prorated. To provide consistency with other assumptions such as that of demand leveling by year fifty, we assume that this source of technological change will gradually reduce to zero at year fifty also.

One form of technological change not effectively incorporated into the model is that which is wood extending and makes forests and species with previously limited industrial uses now industrially usable. Such technological change can be viewed as shifting the economic supply curve outward, independent of genetic improvements, more rapid growth, and the like. To the extent that the absence of this feature biases the TSM, the model will tend to underestimate total harvests and overestimate real prices.

Appendix M
The Yield Function

The yield function plays an integral role in the TSM because it determines yield per hectare and, along with the interest rate and the path of stumpage price, the optimal rotation period. The yield of merchantable volume in cubic meters per hectare is a function of the age of the stand, the level of management practices (regeneration input) applied to the hectare, and the level of technology associated with the management practices.

The equation used has the following characteristics. The parameters for the part that relates age to yield, $q^2(\text{age})$, are estimated from yield tables for fully stocked stands for all land classes except the emerging region, the Nordic land classes, and the Asia-Pacific region. The parameters for these land classes have been selected to yield specific characteristics, as described later. The aging portion fits the data from the yield tables very well. The R-squared statistic exceeded 0.99 for all empirically derived aging yield functions.

The natural stand is a portion (c^2), determined from inventory data, of the natural fully stocked stand. The parameter c^2 has been selected to induce conformity between the numbers calculated by the yield equation for zero regeneration input and the inventory data. The portion of the yield function that incorporates the effects of management practices, z, also incorporates the effects of embodied biotechnological change (btc), $q^1(z, btl)$, where btl is the level of biotechnology. The equation first exhibits the diminishing marginal product of the regeneration input; second, generates a fully stocked stand with a $500 expenditure on the regeneration input; and, third, incorporates the feature that the productivity of the regeneration input is proportional to the level of biotechnology, where that level is initially 1 and grows at the rate specified in the scenario examined. These features are encompassed in

$$q = q^1(z, btl) \cdot q^2(\text{age})$$

$$q^1(z, btl) = \begin{cases} (z + 1)^{c^1 + btc} & \text{for } z \le \bar{z} \\ (\bar{z} + 1)^{c^1 + btc} + a \cdot z + b \cdot z^2 + c & \text{for } z > \bar{z} \end{cases}$$

where

$$btc = btc(btl)$$

$$q^2(\text{age}) = \begin{cases} c^2 \cdot \exp(c^3 + c^4)/(\text{age} - c^5) & \text{for age} > c^5 \\ 0 & \text{for age} \le c^5 \end{cases}$$

where $\bar{z}$ is the level of regeneration input at which the marginal product of the regeneration input switches from a power function to a linear function, and *btl* is biotechnological change. This switch is initiated to limit the productivity of the regeneration input at high levels of input.

The parameters c^3, c^4, and c^5 were estimated from yield table data with the following exceptions. In the emerging region they were selected to yield a Faustmann rotation of 15 years at a 6 percent interest rate, a mean annual increment of 20 m^3 at age 15, and a culmination of mean annual increment at about age 18. For the Asia-Pacific region the equation for q^2(age) was modified slightly by squaring age. This was done to make the yield less sensitive to age so that the function would fit the meager data more closely (see Adiwiyoto [1984], Keil [1978], and Ross [1983]). The Asia-Pacific region's parameters were selected to give a Faustmann rotation of forty years and a mean annual increment of 1 at that age, with c^5 equal to zero. The average yield data for the two Nordic land classes were used in an "eyeball fitting" of the equation to the data, with c^5 equal to zero. The parameter c^2 was calculated as the ratio of average merchantable volume per hectare from inventory data to merchantable volume per hectare for a natural fully stocked stand. Finally, c^1 was calculated from the equation

$$(z + 1)^{c^1} \cdot c^2 = 1$$

with z equal to 500. Thus, with z equal to zero the result is a naturally regenerated stand, and with z equal to 500 the result is a fully stocked stand. The coefficients a, b, and c in the quadratic portion of q^1 are calculated to satisfy the following requirements: (1) the marginal product function for the regeneration input is continuous for all positive z but has a kink at $\bar{z}$, and (2) the marginal product decreases to zero at z equals 850.

Biotechnological change is embodied through *btc* in the q^1 function. At the initial (starting) level of technology, $btl = 1$ and $btc = 0$. Then over time the level of technology grows as specified in the scenario, yielding *btl* greater than 1 and *btc* greater than zero. We use $(z_f + 1)^{btc} = btl$ to calculate *btc*, where z_f is the level of regeneration input that results in a fully stocked stand. Thus over time the productivity of the regeneration input increases because of biotechnological change, with the fully stocked level of regeneration input receiving the full effect of such change.

The parameter values for the twenty-two land classes are listed in table M-1.

Table M-1. Yield Function Parameter Values

Land class	c^1	c^2	c^3	c^4	c^5	$\bar{z}$
1	0.1474	0.4	6.520	−6.5889	6.9	650
2	0.1115	0.5	7.320	−56.9222	10	650
3	0.1115	0.5	7.320	−56.9222	10	650
4	0.1115	0.5	6.770	−76.2117	0	650
5	0.1115	0.5	6.770	−76.2117	0	650
6	0.1115	0.5	6.204	−37.0169	18	650
7	0.1115	0.5	6.204	−37.0169	18	650
8	0.050	0.3	7.697	−90.0	0	650
9	0.050	0.5	7.697	−135.0	0	650
10	0.1474	0.4	6.1367	−10.2576	7	650
11	0.1474	0.4	6.1367	−10.2576	7	650
12	0.1474	0.4	6.0739	−15.3563	6	650
13	0.1474	0.4	6.0739	−15.3563	6	650
14	0.1115	0.5	5.9272	−24.0982	3	650
15	0.1115	0.5	5.9272	−24.0982	3	650
16	0.050	0.5	7.697	−135.0	0	650
17	0.050	0.5	7.697	−135.0	0	650
18	0.1115	0.5	6.135	−28.4649	14	650
19	0.1115	0.5	6.135	−28.4649	14	650
20	0.1115	0.5	5.283	−15.1954	15	650
21	0.1115	0.5	5.283	−15.1954	15	650
22	0.0	1.0	4.689	−1600.0	0	650

REFERENCES

Adiwiyoto, Bambang, P. 1984. "A Discrete Time Optimal Control Model for Optimization in Plywood Industry: A Case Study of Southeast Asia," Ph.D. dissertation (Logan, Utah, Utah State University).

Keil, C. E. M. 1978. "Logging and Log Processing in Indonesia—Forestry Sector Input Study for Basic Economic Work," World Bank Working Document (Washington, D.C., IBRD).

Ross, M. S. 1983. "How Much Production Forest Land Should Indonesia Plan to Keep?" Paper presented at the First ASEAN Forestry Congress, October 10–16, Manila, Philippines.

Appendix N
Regeneration Functions

Each region has a unique yield function based on the conditions and species of the region (see appendix M). The yield function has two components: the regeneration input and the age of the stand. The relationship between age and yield was derived from the data for fully stocked stands. This relationship was adjusted on the basis of regeneration input or lack thereof. An injection of $500 per hectare (at 1980 dollars and exchange rate) of regeneration input resulted in yields equal to those of a fully stocked natural stand. In addition, a zero expenditure generates a natural stand, which is some proportion of the fully stocked stand.

On the basis of data of the various regions an average stand condition for each region was estimated as a percentage of fully stocked. For example, in the Pacific Northwest the average natural stand, that with zero regeneration input, was treated as 50 percent of fully stocked (Berg, 1974). This percentage is based on the relation between a fully stocked stand and the actual inventories given in Forest Service data. Thus in this view the major purpose of initial investments in regeneration is to ensure that the forest is fully stocked with desired species. Subsequent investments can be viewed as promoting growth in excess of that obtainable from a fully stocked stand (see Hyde, 1980, pp. 93–114).

The same procedure was used for each of the other regions with the exception of the Asia-Pacific and emerging regions. For the Asia-Pacific region a yield function was constructed that gave rotations at forty years equal to the volumes expected from the unevenly aged, selectively logged forests. For the emerging region, which consists of exotic, intensively managed stands, the base was a fully stocked stand that had the equivalent of $500 per hectare of investment (at 1980 dollars and exchange rate) in stand establishment and maintenance.

A regeneration function was estimated that related increases in stand stocking with the dollar value of investments in regeneration. This function was developed using data found in Hyde (1980) and Curtis and coauthors (1982). Where data existed and doing so was deemed appropriate, the function was modified to capture the conditions of other regions where information was available. Where data were absent, a regeneration function of the estimated proportions was adapted to the regional yield function. The basic function generated a fully stocked stand with an investment of $500 per hectare. Investments in excess of $500 per hectare were allowed but were limited to diminishing returns that provided no additional output after $850 per hectare. For investment levels that were below $500 per hectare, yields were reduced continuously, falling to the levels of unmanaged stands when the investment level reached zero.

For the industrial plantations of the emerging region, the approach was similar to that discussed above. Because it was recognized that in most cases in the emerging region some threshold level of investment is necessary to generate any production from these forests, the model was utilized on the basis of a threshold level below which no production was forthcoming. However, the result of this was that the emerging region moved in and out of production in a discontinuous fashion. This result was viewed as unrealistic, as the emerging region is really an aggregation of numerous different subregions with somewhat different local conditions. It would be expected instead that although the various subregions might move in and out of production as marginal conditions changed, the entire aggregated region would be unlikely to do so. Therefore, it was determined that an aggregate regional regeneration function of the general type used for the other regions would be the best proxy for real-world marginal adjustment in the emerging region's aggregate harvest levels.

For the Asia-Pacific region a yield function was developed that provided for a rotation period of forty years and a harvest volume equivalent to that expected under the present selective harvesting regime with a harvesting cycle of thirty-five to forty years. In this manner an even-aged, clearcut management regime and yield function were adopted to capture the selective logging and harvesting approach common in that region (see Adiwiyoto, 1984).

REFERENCES

Adiwiyoto, Bambang P. 1984. "A Discrete Time Optimal Control Model for Optimization in Plywood Industry: A Case Study of Southeast Asia," Ph.D. dissertation (Logan, Utah, Utah State University).

Berg, Alan B. 1974. *Managing Young Forests in the Douglas-Fir Region* vol. 4 (Corvallis, Oregon State University, School of Forestry).

Curtis, Robert O., Gary W. Clendenen, Donald Reukema, and Donald J. DeMurs. 1982. "Yield Tables for Managed Stands of Coast Douglas Fir," USDA-Forest Service General Technical Report no. PNW-135 (Washington, D.C., USDA).

Hyde, William. 1980. *Timber Supply, Land Allocation, and Economic Efficiency* (Baltimore, Md., Johns Hopkins University Press for Resources for the Future).

Appendix O
The Formal Timber Supply Model

This appendix provides technical support for chapter 7. We derive equations (7-14), (7-15), and (7-16), which were used extensively in chapter 7 but not rigorously derived, and generate and discuss our solution algorithm. In chapter 7 the discrete time optimal control (DTOC) theory was used to identify the equalities and inequalities that exist at the optimum. This information is used in the solution algorithm and in theoretical discussions of the economics of forestry. The necessary conditions of the optimal control model are manipulated to identify a difference equation problem that has initial and terminal conditions. The solution to this boundary value problem is identified using a search routine that repetitively and numerically evaluates (shoots) the difference equations. The solution is the trajectory that satisfies the initial and terminal conditions.

In chapter 7 we described the model, identified its equations, stated the discrete time maximum principle that is used, and identified the first-order necessary conditions. We will use that material without restating it here. We will first generate the analytic solution for the transition period and for the stationary state. This will provide the technical support for equations (7-14), (7-15), and (7-16) as well as the foundation for our solution algorithm. Then we will discuss the difference equations that we solve numerically to implement the model, and finally describe our shooting method algorithm.

THE ANALYTIC SOLUTION

The Transition Period

Equations (7-7a), (7-7b), (7-11a), (7-11b), and (7-13a through f) are the equations to be solved. Equations (7-7a) and (7-7b) identify the method for calculating the values of state variables (hectares of forest by age group and stock of regeneration investment) in each year, given the values for the control variables in each year. These equations move the state variables forward over time.

Equations (7-11a) and (7-11b) identify the method for calculating the costate variables (shadow values of the state variables) in each year, given the values of the control variables in each year. This procedure calculates the costate variables starting at year J (the last time period) and moving backward through time to the present (year $j = 0$). Finally, equations (7-13a through f) identify the method of finding the values of control variables in each year.

We now examine the information in these equations, starting with the costate difference equations (7-11a) and (7-11b). If we define

$$\alpha_{hj} = [D(Q_j) - c_h'(Q_{hj})]U_{hj}q_{hj} - u_{hj}w_{hj}p_{wh} \tag{1}$$

we can write

$$\lambda_{hj} = \rho dS_J^* / dx_{hJ} \tag{2a}$$

$$\lambda_{hj} = \rho[\alpha_{hj} + (A + BU_{hj})'\lambda_{h,j+1} \qquad (j = 1, \ldots, J - 1) \tag{2b}$$

$$\begin{aligned}\lambda_{h1} = \rho[\alpha_{h1} &+ (A + BU_{h1})'\alpha_{h2} + \rho^2(A + BU_{h1})'(A + BU_{h2})'\alpha_{h3} \\ &+ \ldots + \rho^{J-2}(A + BU_{h1})'(A + BU_{h2})' \ldots (A + BU_{h,J-2})'\alpha_{h,J-1} \\ &+ \rho^{J-1}(A + BU_{h1})'(A + BU_{h2})' \ldots (A + BU_{h,J-1})'(dS_J^*/dx_{hJ})]\end{aligned} \tag{2c}$$

Equation (2c) is useful in calculating values for the stationary state, and equation (2b) is useful in calculating the values of λ_{hj}, which are evaluated moving from year J backward to year one. In the calculations the values for λ_{hJ} are those from the stationary state.

Because the oldest trees will be harvested first, there will be at most one u_{hij} that is not either zero or 1. Let k be age of the youngest age group harvested in year j and let $u_{hij} = 1$ for $i > k$. This explains that we are harvesting in year j all existing trees for which the age is greater than k. We may have zero trees in the age group that are older than age k, in which case the only value of this assumption is in interpreting the equation. We can express u_{hj} as

$$u_j = \begin{bmatrix} 0 \\ \cdot \\ \cdot \\ \cdot \\ 0 \\ u_{kj} \\ 1 \\ \cdot \\ \cdot \\ \cdot \\ \cdot \\ 1 \end{bmatrix} \tag{3a}$$

and $(A + BU_j^*)'$ as

$$(A + BU_j)' = \begin{bmatrix} 0 & 1 & . & . & . & 0 & 0 & . & . & . & . & 0 \\ 0 & 0 & 1 & . & . & 0 & . & . & . & . & . & . \\ . & . & . & . & . & . & . & . & . & . & . & . \\ . & . & . & . & . & . & . & . & . & . & . & . \\ 0 & . & . & . & . & 1 & 0 & . & . & . & . & . \\ u_{kj} & 0 & . & . & . & (1-u_{kj}) & . & . & . & . & . & . \\ 1 & . & . & . & . & . & . & . & . & . & . & . \\ . & . & . & . & . & . & . & . & . & . & . & . \\ . & . & . & . & . & . & . & . & . & . & . & . \\ . & . & . & . & . & . & . & . & . & . & . & 0 \\ 1 & 0 & . & . & . & 0 & 0 & . & . & . & . & 0 \end{bmatrix} \tag{3b}$$

Thus

$$\alpha_{hj} = \begin{bmatrix} 0 \\ . \\ . \\ 0 \\ [p_{hj}q_{hk} - p_{wh}w_{hj}]u_{hkj} \\ p_{hj}q_{h,k+1} - p_{wh}w_{hj} \\ . \quad . \qquad . \quad . \\ . \quad . \qquad . \quad . \\ p_{hj}q_{hM} - p_{wh}w_{hj} \end{bmatrix}$$

where p_{hj} is net price or stumpage price of timber for land class h. Stumpage price is equal to the market price of timber, $D(Q_j)$, minus the marginal harvesting, accessing, and transportation cost of timber, $c_h'(Q_{hj})$. Using these results we get

$$\lambda_{hij} = \rho\lambda_{h,i+1,j+1} \qquad \text{for } i < k \tag{4a}$$

$$\lambda_{hkj} = \rho[(\lambda_{h1,j+1} + p_{hj}q_{hk} - p_{wh}w_{hj})u_{hkj} + \lambda_{h,k+1,j+1}(1 - u_{hkj})] \tag{4b}$$

$$\lambda_{hij} = \rho(\lambda_{h1,j+1} + p_{hj}q_{hi} - p_{wh}w_{hj}) \qquad \text{for } i > k \tag{4c}$$

because $u_{hij} = 1$ for $i > k$. Equation (4c) is the source of equation (7-15); however, note that in the optimal time path if the harvest of an age group is spread over more than one year, then equation (4b) is the correct source for (7-15). The two equations, however, yield the same result in the optimal time path because the timber in this age group will have the same value in each year that it is harvested. This implies that its shadow value λ will also have the same value in each year that the age group is harvested.

We can now use similar procedures to examine equation (7-11b). If we define

$$\beta_{hj} = p_{hj}(dq_h/dz_{hj})X_{hj}u_{hj} \tag{5}$$

we can write

$$\psi_{hJ} = \rho \partial S_J^* / \partial z_{hJ} \tag{6a}$$
$$\psi_{hj} = \rho(\beta_{hj} + A'\psi_{h,j+1}) \quad \text{for } j < J \tag{6b}$$
$$\psi_{h1} = \rho[\beta_{h1} + \rho A'\beta_{h2} + (\rho A')^2\beta_{h3} + \ldots + (\rho A')^{J-2}\beta_{h,J-1} + (\rho A')^{J-1}(dS_J^*/dz_{hJ})] \tag{6c}$$

Equation (6c) is useful in calculating values of these costate variables in the stationary state, and equation (6b) is useful in calculating these values in the transition period. The procedure uses $\psi_{h,j+1}$ and moves backward through time from year J to year one. The elements of β_{hj} are

$$\beta_{hij} = 0 \quad \text{for } i < k \tag{7a}$$
$$\beta_{hkj} = p_{hj}(\partial q_{hk}/\partial z_{hkj})x_{hkj}u_{hkj} \tag{7b}$$
$$\beta_{hij} = p_{hj}(\partial q_{hi}/\partial z_{hij})x_{hij} \quad \text{for } i > k \tag{7c}$$

This last equation will equal zero for most values of i because x_{hij} will equal zero for most i greater than k. Thus

$$\psi_{hij} = \rho\psi_{h,i+1,j+1} \quad \text{for } i < k \tag{8a}$$
$$\psi_{hkj} = \rho[p_{hj}(\partial q_{hk}/\partial z_{hkj})x_{hkj}u_{hkj} + \psi_{h,k+1,j+1}] \tag{8b}$$
$$\psi_{hij} = \rho[p_{hj}(\partial q_{hi}/\partial z_{hij})x_{hij} + \psi_{h,i+1,j+1}] \quad \text{for } i > k \tag{8c}$$

To examine equation (7-13a) we use

$$X'_{hj}B\lambda_{h,j+1} = \begin{bmatrix} x_{h1j} & -x_{h1j} & \cdot & \cdot & \cdot & 0 \\ x_{h2j} & 0 & -x_{h2j} & \cdot & \cdot & 0 \\ \cdot & \cdot & \cdot & \cdot & \cdot & \cdot \\ \cdot & \cdot & \cdot & \cdot & \cdot & \cdot \\ \cdot & \cdot & \cdot & \cdot & \cdot & 0 \\ x_{h,M-1,j} & \cdot & \cdot & \cdot & 0 & -x_{h,M-1,j} \\ x_{h,M,j} & 0 & \cdot & \cdot & \cdot & 0 \end{bmatrix} \begin{bmatrix} \lambda_{h1,j+1} \\ \lambda_{h2,j+1} \\ \cdot \\ \cdot \\ \cdot \\ \cdot \\ \lambda_{hM,j+1} \end{bmatrix}$$

Thus the elements of dL_j^H/du_{hj} in equation (7-13a) can be written

$$p_{hj}x_{hij}q_{hi} - x_{hij}p_{wh}w_{hj} + x_{hij}(\lambda_{h1,j+1} - \lambda_{h,i+1,j+1}) - \zeta_{hij} \leq 0 \quad \text{(for all } h \text{ and } i) \tag{9}$$

Equation (9) is the source of equation (7-14) and gives the marginal net surplus (net shadow value) of harvesting hectares of trees by age group and land class. If (9) sans ζ_{hij} is positive (negative) for some hectare of trees in year j, then they will (not) be harvested in that year, because doing so adds more (less) to net surplus than letting them age another year. We use equation (9) for the marginal age group to generate a difference equation for $p_{h,j+1}$, hence $D_{j+1}(Q_{j+1})$, and we harvest all hectares of trees with positive marginal net surplus. To derive the difference equation we substitute equation (4c) into (9), letting i equal k, and solve for $p_{h,j+1}$. This yields

$$p_{h,j+1} = [p_{hj}q_{hk} - p_{wh}(w_{hj} - \rho w_{h,j+1}) + \lambda_{h1,j+1} - \rho\lambda_{h1,j+2} - \zeta_{hkj}/x_{hkj}]/\rho q_{h,k+1} \quad (10)$$

This equation is the source of equation (7-16).

The solution to this DTOC model is the time paths of state variables, control variables, and costate variables such that the laws of motion for state variables, the laws of motion of costate variables, the difference equation for the net price of timber (stumpage price) given in equation (10), and the remaining first-order conditions (i.e., equations [7-13b through f]) are simultaneously satisfied. These form a two-point boundary value difference equation problem.

Stationary State

One of the objectives of this section is to identify the vectors of costate variables for the terminal time period. We assume that the terminal time period is sufficiently distant that it will be in the stationary state; hence we identify these vectors from the solution for the stationary state. To find this solution we let all years of the current analysis be in the stationary state; and to prevent further complications of the notation, we let the time periods range from 0 to J as above.

The solution for the stationary state is the solution to the equations discussed earlier for the transition period, with the addition that each year is alike. We let k be the age of the youngest age group harvested in land class h and let $u_{hkj} = 1$. In addition, for solution vectors $u_{hj} = u_{h,j+1}$, $w_{hj} = w_{h,j+1}$, $\alpha_{hj} = \alpha_{h,j+1}$, and $\beta_{hj} = \beta_{h,j+1}$ for all j and $j + 1$ in the stationary state. These equalities yield simplifications of the laws of motion of the costate variables and of the necessary conditions.

Equation (2c) becomes

$$\lambda_{hj} = \rho\{I + \rho(A + BU_{hj})' + [\rho(A + BU_{hj})']^2 + \ldots + [\rho(A + BU_{hj})']^{J-2}\}\alpha_{hj} + \rho[\rho(A + BU_{hj})']^{J-1}(dS^*_{hJ}/dx_{hJ}) \quad (11)$$

where all years are in the stationary state. Given $0 \leq \rho \leq 1$ and J is very large, we can ignore the last term in this equation. The evaluation of equation (11) yields

$$\lambda_{h1} = \begin{bmatrix} \rho^{k}\Delta\alpha_{hkj} \\ \rho^{k-1}\Delta\alpha_{hkj} \\ \cdot \\ \cdot \\ \rho\Delta\alpha_{hkj} \\ \rho[(\Delta - 1)\alpha_{hkj} + \alpha_{h,k+1,j}] \\ \cdot \\ \cdot \\ \rho[(\Delta - 1)\alpha_{hkj} + \alpha_{hMj}] \end{bmatrix} \quad (12)$$

where α_{hkj} are defined in equation (3c) and

$$\Delta = 1 + \rho^k + \rho^{2k} + \ldots \quad (13)$$

Turning our attention now to ψ_{h1}, we write equation (6c) for the stationary state as

$$\psi_{h1} = \rho[I + \rho A' + (\rho A')^2 + \ldots + (\rho A')^{J-2}]\beta_{hj} + \rho(\rho A')^{J-1}(dS^*_{hJ}/dz_{hj}) \quad (14)$$

As above, the last term will be dropped. In land class h there are no trees older than k because we let $u_{hkj} = 1$. This implies that the only nonzero element in β_{hj} is $\beta_{hkj} = p_{hj}(\partial q_{hk}/\partial z_{hkj})x_{hkj}$. Therefore

$$\psi_{h1} = \begin{bmatrix} \rho^k \beta_{hkj} \\ \rho^{k-1}\beta_{hkj} \\ \cdot \\ \cdot \\ \rho\beta_{hkj} \\ 0 \\ \cdot \\ \cdot \\ 0 \end{bmatrix} \quad (15)$$

Combining equations (9) and (12) and combining equations (7-13c) and (15) yields a set of equations that determine the solution stationary state rotation period and regeneration input in each land class. These are

$$[p_{hj}q_{hk} - p_{wh}w_{hj}](1 - \rho)\Delta - p_{hj}(q_{hk} - q_{h,k-1}) < 0 \quad (16a)$$

$$[p_{hj}q_{hk} - p_{wh}w_{hj}](1 - \rho)\Delta - \rho p_{hj}(q_{h,k+1} - q_{h,k}) \geq 0 \quad (16b)$$

and

$$\rho^k p_{hj}(\partial q_{hk}/\partial w_{hj}) - p_{wh} = 0 \quad (16c)$$

To solve for a numerical solution our algorithm uses a search routine to solve these equations.

THE SOLUTION ALGORITHM

Our solution algorithm solves for the optimal values of the control (choice) variables in the transition period. These time paths generate an evolution of the state variables (for example, hectares of trees by age group) from their initial values to those in the stationary state. The algorithm uses a shooting (binary search) method to solve a constrained difference equation problem. The difference equations are the laws of motions of the state variables, equations (7-7a) and (7-7b); the laws of motions of costate variables, equations (2b) and (6b); and equation (10), which is derived from the first two necessary conditions, equations (7-13a) and (7-13b). These form a two-point boundary value difference equation problem that is to be solved subject to the remaining first-order conditions, equations (7-13c through f). The boundary values are determined by the initial and terminal conditions.

These difference equations with their initial conditions have a set of solutions, one for each value of market price of timber in the first time period. The shooting method is a search for the member of this set that satisfies the terminal conditions. This search is carried out by starting with an arbitrary element from the set of solutions and systematically eliminating solutions that do not satisfy the terminal conditions until a satisfactory solution is found.

The Procedure

The algorithm first calculates the stationary state values of all variables, recognizing the initial conditions and timber yield functions.The optimum stationary state rotation periods and regeneration investment for all land classes are calculated from equations (16a), (16b), and (16c) using a search routine.

The problem for the transition period is solved using a three-step procedure. The first finds an initial feasible time path of control and state variables using an arbitrary but reasonable time path of the costate variables, which are calculated using a backward-moving difference equation that requires a feasible time path of the state variables. These time paths are found by solving the difference equation problem using a shooting (binary search) method described below (Lyon and Sedjo, 1986).

Using the resulting time profiles of the state variables, the next step is to calculate the costate variables moving from long-run equilibrium backward to year one by using the backward-moving laws of motion, equations (2b) and (6b).

On the basis of the results of the second step, we now use the shooting method to solve for the time paths of the control and state variables. The only difference between this step and the first step is that calculated time paths of the costate variables are used instead of arbitrary ones. In this we solve the difference equations (10), (7-7a), and (7-7b) subject to a set of necessary conditions, equations (7-13c through f), to determine the new time profiles of state and control variables from the initial conditions to a long-run equilibrium. These last two steps, which calculate the costate variables in the second step and determine the new time profiles of state and control variables in the third step, can be repeated until a satisfactory solution is determined.

The Shooting (Binary Search) Method

The shooting (binary search) method as implemented here is a search for a particular element of a set of solutions to the equations and initial conditions. This element is the one that satisfies the terminal conditions. To start the search, an arbitrary value for the market price of timber in the initial year is selected, and the system is shot—that is, the equations are numerically evaluated to yield a trajectory (time path) for the market price and the state variables. If the trajectory misses the target (terminal condition), the stumpage price in the initial time period is adjusted up or down, depending on the direction of the miss, and the process is repeated.

The level of per-hectare regeneration input can be calculated for the entire transition period using equations (7-13c) and (7-13d) from the first-order conditions and the time path of the costate variable for z_h, ψ_h. The first step in calculating the harvest levels is to select arbitrarily the market price of timber for the first time period (j = 0). The mathematical relationships given by the first-order conditions, equations (7-13a), (7-13b), (7-13e), and (7-13f), and the time paths of the costate variables for x_h, λ_h are then used to calculate the static market-clearing values of the timber harvests by land class. This static problem in a particular year can be solved by recognizing the timber demand function and the harvesting and transporting cost functions. The demand function and the market price determine the quantity to be harvested, Q_0. On the basis of the information of the forest initial conditions and the timber yield function, we harvest the oldest-age group of trees first within a land class and harvest first from the land class containing the age group with the highest net shadow value of harvested timber ($\partial L_j^H / \partial u_{hkj}$, where k is the age of the oldest age group in land class h in year j, equation [9]). This net shadow value on a per-hectare basis is stumpage price times volume per hectare minus the opportunity cost of harvesting the trees in the current year. The harvesting continues from one exhausted age group to the age group with highest net shadow value until Q_0 m^3 of timber have been harvested. This determines by land class, hectares of trees harvested and regenerated, youngest age group harvested, and the level of the regeneration input applied to the regenerated hectares. The last age group harvested, called the marginal age group, is used to calculate the market price in the next time period. If the age group is not completely exhausted, equation (10) can be used to calculate the stumpage price of this age group in the next year because ζ_{hkj} will equal zero and the other variables are known.

The market price of timber in the next year is then calculated from the stumpage price using the definition of stumpage price. If, however, the marginal age group is exhausted exactly, a restart procedure (discussed later) is used to determine the value of ζ_{hkj}. We then age the remaining trees one year by implementing the law of motion for hectares of trees by age group. Following this we move forward one year and statically solve the same problem for this year.

This process of statically solving the first-order conditions and dynamically evaluating the difference equations yields a time profile of the market price of timber and state variables. The process is iterated until a particular time period when we realize that the price in the initial time period is too low or too high, or until the time horizon, the arbitrarily selected end of the transition period, is reached. If the time profile of price over-utilizes (under-utilizes) the forest resources, the price in the initial time period is too low (high). In this case the process needs to be repeated. A new price in the initial time period is calculated by selecting the midpoint between the lowest price that is known to be high and the highest price that is known to be low.

The iterative process stops when the difference between the two levels of price that bracket the optimum level becomes sufficiently small that continued iterations yield no significant information.

When the marginal age group is exhausted exactly, equation (10) contains a Lagrangean multiplier ζ_{hkj} of unknown value. This multiplier is the shadow value of the constraint $u_{hkj} \leq 1$, and ζ_{hkj}/x_{hkj} is the per-hectare shadow value of this constraint. Equations (7-13e) and (7-13f) reveal that $\zeta_{hkj} = 0$ if $u_{hkj} < 1$; that is, the harvest in year j does not exhaust the age group, and $\zeta_{hkj} \geq 0$ if $u_{hkj} = 1$. If two age groups in land class h are harvested in some time period j, the older one $(k + 1)$ that is completely harvested has $\zeta_{h,k+1,j}$ greater than zero, and $\zeta_{h,k+1,j}/x_{h,k+1,j}$ will identify the difference in per-hectare value of $k + 1$ and k age group trees.

When the harvest in a time period exhausts exactly the youngest age group cut, ζ_{hkj} will be greater than or equal to zero. Its function at this point is to smooth the discontinuity caused by using discrete age groups. If the solution ζ_{hkj} is greater than zero, the solution market price in the next year is less than the one calculated using equation (10) and this marginal age group with ζ_{hkj} set equal to zero, and the solution market price next year is greater than or equal to the one calculated using the next candidate for the marginal age group. In implementing the binary search, this omission will cause for this situation a fluctuation between these two as the marginal age group.

When ζ_{hkj} is nonzero in year j, the solution price in year $j + 1$ can be found by restarting the binary search in year $j + 1$ and searching over the range of interval of prices identified earlier. Year $j + 1$ is treated as the initial year, and the initial conditions for this new problem are the conditions left after year j.

REFERENCES

Lyon, Kenneth S., and Roger A. Sedjo. 1982. "Discrete Optimal Control Algorithm for Analysis of Long-Run Timber Supply," Working Paper no. 96 (Washington, D.C., Resources for the Future).

Lyon, Kenneth S., and Roger A. Sedjo. 1983. "An Optimal Control Theory Model to Estimate the Regional Long-Term Supply of Timber," *Forest Science* vol. 29, no. 4, pp. 798–812.

Lyon, Kenneth S., and Roger S. Sedjo. 1986. "Binary Search SPOC: An Optimal Control Theory Version of ECHO," *Forest Science* vol. 32, no. 3, pp. 576–584.

Glossary of Selected Terms

ALTERNATIVE SCENARIOS: various future situations, each having different but quite conceivable underlying assumptions, that are examined for their effects on the long-term timber supply of industrial wood. They include (a) alternative rates of growth in long-term worldwide demand; (b) alternative assumptions about the world's exchange rate structure; (c) alternative levels of newly established forest plantations in emerging producer regions; (d) an assumption of increased timber harvests in the Soviet Union; and (e) an assumption of a fiscal "shock" on one major producer, the United States, in the form of higher taxes on forestry investment activities.

BASE-CASE SCENARIO: the authors' best judgment as to the future course of prices and harvests for industrial wood in the seven **responsive regions** of the timber-producing world. This scenario is most likely in a forecasting sense, given what the authors view to be the most probable assumptions about future events, the data used in their **Timber Supply Model,** and the conceptual consistency imposed on the projections by the model.

CONTROL THEORY APPROACH: the methodological approach utilized by the **Timber Supply Model** developed by Sedjo and Lyon. This approach explicitly introduces initial conditions (such as area of forest by age group and **land class**) and laws of motion describing rules that govern the **world timber system** over time. Unlike the linear programming approach, with the control theory approach the changing age and volume conditions of the forest are constantly monitored and updated so that management decisions can recognize explicitly the changing state of the forest.

EMERGING REGION: one of the seven **responsive regions** modeled in the **Timber Supply Model.** This region is a composite of tropical and semitropical subregions that have established industrial forest plantations of nonindigenous species in the past twenty-five or so years. Comprising temperate areas of the Southern Hemisphere and the Iberian Peninsula, it includes parts of Brazil, Chile, Venezuela, Australia, New Zealand, South Africa, Spain, and Portugal. Recent industrial forest plantation activity in the emerging region suggests the possibility of large volumes of future production.

LAND CLASS: one of the twenty-two basic units of production into which the seven **responsive regions** of the **Timber Supply Model** are disaggregated. Each land class is treated as uniform, and incorporates physical and biological information to develop a unique production function.

NONRESPONSIVE REGIONS: the regions that are not modeled in the **Timber Supply Model.** These regions—the Soviet Union, non-Nordic Europe, and the rest of the world not included in the **responsive regions**—lack substantial exploitable stands of timber resources or lack available data, or their timber production is unlikely to be responsive to market forces and incentives due to the economic-cultural system of the regions. The harvest levels of the three nonresponsive regions are included in the global analysis. However, the harvests are viewed as autonomous and determined independently of the usual economic calculus.

OLD GROWTH FOREST: one of three categories of industrial forest in the **world timber system.** These timber stands are essentially virgin, being relatively undisturbed by human activity, and often have very large wood volume.

PLANTATION FORESTS: one of three categories of industrial forest in the **world timber system.** These planted and intensively managed forests are often established on land that had not previously had important industrial forests. They sometimes consist of native tree species, although many have introduced exotic (nonindigenous) species.

RESPONSIVE REGIONS: the seven forest regions, accounting for about one-half of the world's industrial wood, that are individually modeled and incorporated into the mathematical **Timber Supply Model** (TSM) developed by Sedjo and Lyon. They are (1) the U.S. South, (2) U.S. Pacific Northwest, (3) western Canada (British Columbia), (4) eastern Canada, (5) Nordic Europe, (6) Asia-Pacific, and (7) **the emerging region.** The TSM assumes economically optimizing behavior, that is, responsiveness to profit-maximizing incentives, on the part of forest owners and managers in these regions.

SECONDARY AND MANAGED FORESTS: one of three categories of industrial forest in the **world timber system.** Secondary (or second-growth) forests are areas that were logged or from which the trees had otherwise been removed at some earlier time, although not so long ago that the forest is like an old growth forest. Managed forests have replaced earlier, indigenous forests.

SITE CLASS: the categorization of forestland by its potential forest-growing productivity. In the **Timber Supply Model,** each of the twenty-two **land classes** is representative of only one site class. "High site land" has rapid timber growth but is designated by a low number—e.g., site class I has highest growth.

TIMBER SUPPLY MODEL: the principal tool of this study, developed by Sedjo and Lyon to assist in the task of assessing the adequacy of the long-term world timber supply. This model is explicitly designed to examine and project the transition from **old growth forest** to **secondary and managed forests** to intensively managed **plantation forests**—that is, the "forest in transition" in a global context. It uses a **control theory approach** to accomplish this task.

WORLD TIMBER SYSTEM: the combination of all industrial forests, in both **responsive** and **nonresponsive regions,** that constitutes the world's industrial forest resources. The **Timber Supply Model** deals simultaneously with all forms of forest.

Index